NEURAL STEM CELLS AND ADULT NEUROGENESIS

NEURAL STEM CELLS AND ADULT NEUROGENESIS

ARIE S. MOBLEY

ELSEVIER

ACADEMIC PRESS
An imprint of Elsevier

Academic Press is an imprint of Elsevier
125 London Wall, London EC2Y 5AS, United Kingdom
525 B Street, Suite 1650, San Diego, CA 92101, United States
50 Hampshire Street, 5th Floor, Cambridge, MA 02139, United States
The Boulevard, Langford Lane, Kidlington, Oxford OX5 1GB, United Kingdom

Library of Congress Cataloging-in-Publication Data
A catalog record for this book is available from the Library of Congress

British Library Cataloguing-in-Publication Data
A catalogue record for this book is available from the British Library

ISBN: 978-0-12-811014-0

Publisher: Nikki Levi
Acquisition Editor: Natalie Farra
Editorial Project Manager: Kristi Anderson
Production Project Manager: Omer Mukthar
Cover Designer: Mark Rogers

Typeset by SPi Global, India

Contents

II

ADULT NEUROGENESIS

Preface—The Promise of Stem Cell Research

Open a newspaper, turn on your TV or radio, go to a news website, and chances are good that you will find a story about stem cells. While we hear a lot about stem cells, there is a general lack of specific knowledge in the public regarding these important cells. It is broadly recognized that they have the potential to cure diseases, but how?

Imagine your doctor had some cells in a dish that were a genetic match to you. These cells were in a "blank" state and could be used to treat any tissue type in your body. For example, if you had Type I diabetes these cells could be altered to become insulin-producing cells. In another scenario, imagine you need a drug for a heart condition, but there is some question regarding its toxicity. Instead of testing the drug directly on you, the cells in the dish could be tested instead. The effect of the drug at the cellular and molecular level could be assessed to determine whether it would work for you. The scenarios described here can all be accomplished using stem cells.

A few years ago I began teaching at a small liberal arts university in a department for undergraduate neuroscience. The school had a fairly new Neuroscience program and my Chair was open to new ideas for course offerings. I wanted to teach a course on stem cells and adult neurogenesis and began looking for books on the topic. While I found many books on stem cells, they were either too broadly or narrowly focused, or too advanced for an undergraduate course. The first time I taught the course I built all my lectures using primary literature and some online resources, such as https://learn.genetics.utah.edu/content/stemcells/. I received a lot of positive feedback from the students regarding the course. They expressed enthusiasm for the topics covered, including discussions of the scenarios in the paragraph above, and an eagerness to learn more. The lack of an appropriate textbook motivated me to write one aimed at undergraduate neuroscience students. The goal of the book is to help students appreciate the potential, and understand the limitations of stem cells, while providing a basic knowledge of stem cell physiology. The next time they come across stem cells in the news, they can evaluate the information based on a foundation of stem cell knowledge gained from this book. The focus on adult neurogenesis in the brain is due to my own research interests that arose during my postdoctoral work in the laboratory of Dr. Charlie Greer.

Chapters 1–3 of the book introduce stem cell concepts such as self-renewal, multipotency and the stem cell niche. Chapter 4 introduces the reader to induced pluripotent stem cells, including concepts such as reprogramming and transdifferentiation. Chapters 5–8 focus on the behaviors and regulation of neural stem cells as well as their relevance to health, disorders and disease.

With each chapter, the student is introduced to a bit of research history, such as some of the Nobel Prize winners for major stem cell-related discoveries. Most of the chapters introduce one or two signaling pathways that are common to several stem cell types. Where appropriate, mechanisms of neurodevelopment that are essential to studying adult neurogenesis are introduced, such as the regulation of symmetric vs. asymmetric cell division and neural migration. The book also includes discussions on the ethics of embryonic stem cell research, cloning, and stem cell therapies. Finally, modern or commonly used techniques are explained so that students understand the experiments that discovered the knowledge presented in this book.

This book would not be possible without help from many people. Importantly, my students must be thanked for inspiring me to write the book for them. Thank you to Jessica Brann for keeping me motivated and helping to turn my idea into reality. Thank you to Sheralee Tershner for letting me teach whatever I wanted. Thank you to Charlie Greer and Dianna Bartel for providing excellent examples of fantastic scientific writing as well as encouragement and support. Thank you to Carolyn Mobley for her illustration. Thank you to those who reviewed and critiqued drafts: Nathaniel Hartman, Emily Ford, Melissa Lail-Trecker and especially Paula Joh Porter. Finally, thank you to my family for supporting me during the writing process.

Arie S. Mobley

SECTION I

STEM CELLS

CHAPTER

1

History and Ethics of Stem Cells

OUTLINE

LEARNING OBJECTIVES

1. Describe the properties of stem cells.
2. Explain how the term "stem cell" was derived.
3. Discuss some of the early experiments and fields of research that led to a basic understanding of stem cell properties.
4. Define clone and cloning.
5. Discuss the ethical issues surrounding ESC research.

https://doi.org/10.1016/B978-0-12-811014-0.00001-9

INTRODUCTION

Why study stem cells? Stem cell research holds the potential to improve several aspects of human health because of two special properties. When stem cells divide they can produce two new copies of the original, a property called *self-renewal*. Some stem cells are *pluripotent*, meaning they can become any cell type in the body through differentiation (Fig. 1.1). Imagine an unlimited source of any cell type you needed. How stem cells regulate, whether to self-renew or *differentiate* into another cell type, is an important function that scientists studying development and cancer

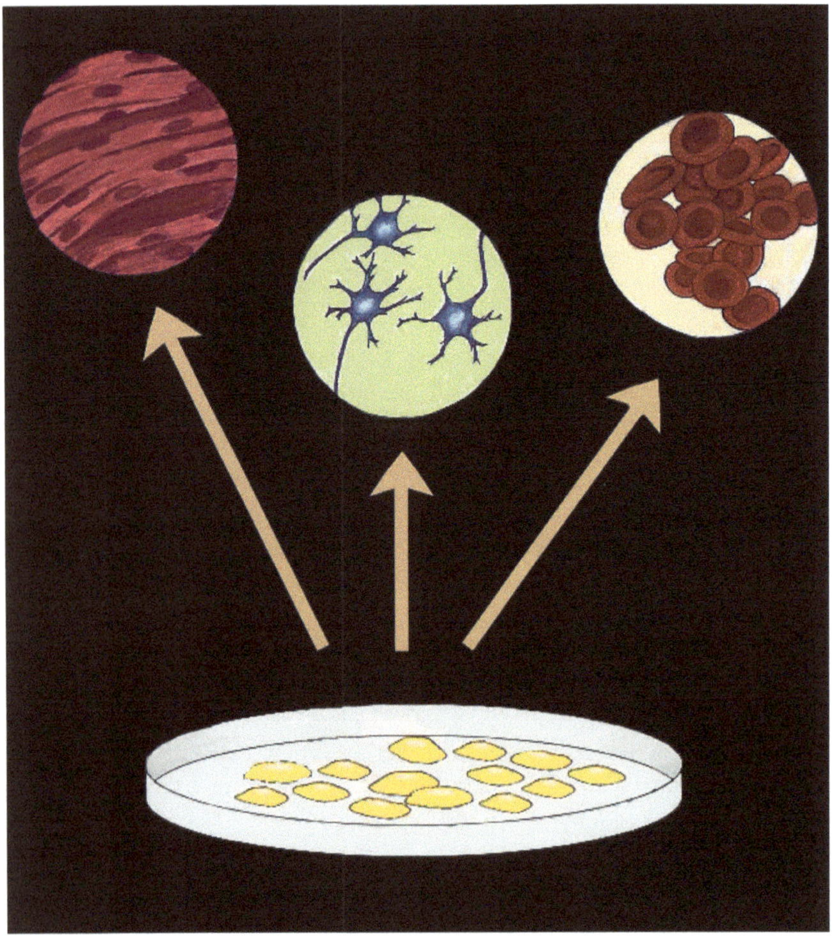

FIG. 1.1 The promise of stem cell research is twofold: an unlimited supply of cells that can become any cell in the body and the potential for therapy or disease prevention, which is still being discovered.

would like to understand. Much is known on how stem cells differentiate into certain cell types and current research is working to harness that knowledge into replacing damaged or degenerating cells. For example, differentiating stem cells into epithelial cells could provide a treatment for burn victims (Chen et al., 2009). Other cell-based therapies are being investigated to replace complicated and expensive whole organ transplants. For instance, injecting stem cells into a failing organ could repair damage resulting in improved organ function and patient health. Finally, differentiated stem cells provide a powerful method to identify drug targets, test drug effects or toxicity levels. This method not only reduces extensive animal testing, but it also speeds up the process for drug approval and the move to market by testing those experimental drugs directly on stem cell-derived human tissue. While questions such as tumor formation and immune response need to be addressed, the potential for improving human health has made stem cells an important area of research.

HISTORY OF STEM CELLS

Etymology of "Stem" Cells

The word "stem" puts us in mind of a plant, perhaps. Why do we call a pluripotent, self-renewing cell a stem cell? The term stem cell can be traced back to German native, Ernst Haeckel. Haeckel earned his medical degree in 1857 but quickly turned to studying the single-celled protozoan group Radiolaria. Darwin's (1859) publication, *On the Origin of the Species* had a profound impact on his studies, leading to his belief that all plants and animals evolved from single-celled organisms (Darwin, 1859). Haeckel drew detailed evolutionary trees to illustrate how he thought various species could be related. In his 1910 fifth edition, *The Evolution of Man: A Popular Scientific Study*, he designed an elaborate branch-like drawing called the Paleontological Tree of the Vertebrates (Haeckel, 1905). In German, he called these drawings "Stammbäume," which translates to family tree or stem tree in English. His work on Radiolaria led to further research on invertebrate development. In 1877 he proposed that the fertilized egg should be called a "Stammzelle" or stem cell. For Haeckel, the word stem referred to single cells as ancestors of multicellular organisms and single cells that give rise to all cells of an organism (Ramalho-Santos and Willenbring, 2007).

In the late 1800s it had become recognized that there was a difference between *germ cells* and *somatic* cells. In 1892 August Weismann published the "Continuity of the Germ-Plasm," which proposed that germ plasm was transmitted from one generation to the next (Weismann, 1892). This reference to genetic material was an important step in understanding the nature of stem cells. That same year Drs. Theodor Boveri and Valentin Häcker independently set out to identify the earliest germ cells in animal embryos, which would presumably carry the germ plasm.

Boveri traced the cell lineages of the nematode *Ascaris*, depicting them as tree diagrams that he also called Stammbäume (Ramalho-Santos and Willenbring, 2007). In the *Ascaris* embryo the earliest mitoses of cells destined to give rise to somatic cells displays an unusual pattern of chromatin fragmentation and loss. In a famous experiment Boveri shook the eggs of two different sea urchin species together, which resulted in the mostly partial hybrid recombination of egg and nucleus. In a small fraction he obtained a complete egg and nucleus hybrid, and demonstrated that only in germ cells with the full complement of chromatin will the embryo develop normally. Boveri proposed that cells that can give rise to a complete embryo be called stem cells (Boveri, 1889).

Meanwhile, Häcker was studying the development of the crustacean *Cyclops*, identifying a large cell that was internalized early in development (Häcker, 1892). Further tracking revealed that these cells later give rise to oocytes in the gonads. Both Häcker and Boveri used the term stem cell to describe cells of the germline lineage.

The term stem cells as well as the concept was advanced around this same time by a very different field of research, hematology. When different white blood cell lineages were discovered hematologists were divided on whether all hematopoietic lineages arose from one common cell or whether there were two types of precursors, one for myeloid cell types and another for lymphoid cell types. A drawing from Artur Pappenheim in 1905 proposed the hematopoietic cell lineages arising from a single cell (Fig. 1.2). The Russian histologist Alexander Maksimov (Maximow) became renowned first, for his Unitarian theory of hematopoiesis (1906), and second, for proposing the term "stem cell" during his delivery of a scientific paper at the Congress of the Hematologic Society in Berlin in 1908 (Maximow, 1909). Maksimov subsequently demonstrated in 1924 that single cells located in the bone marrow could give rise to all blood cell lineages, later confirmed to be *mesenchymal cells* (Maximow, 1924). His announcement at such a prestigious meeting combined with his significant findings cemented the term and its meaning in a large scientific field resulting in its common use.

Early Stem Cell Discoveries

The first use of the term "stem cells" in an abstract/title recorded in PubMed is 1916. Dr. Vera Danchakoff of the Rockefeller Institute for Medical Research published a paper on the differentiation of small cortical cells of the thymus in the "loose mesenchyme" in birds (Danchakoff, 1916). The second mention of stem cells in an abstract/title is in 1932, again from

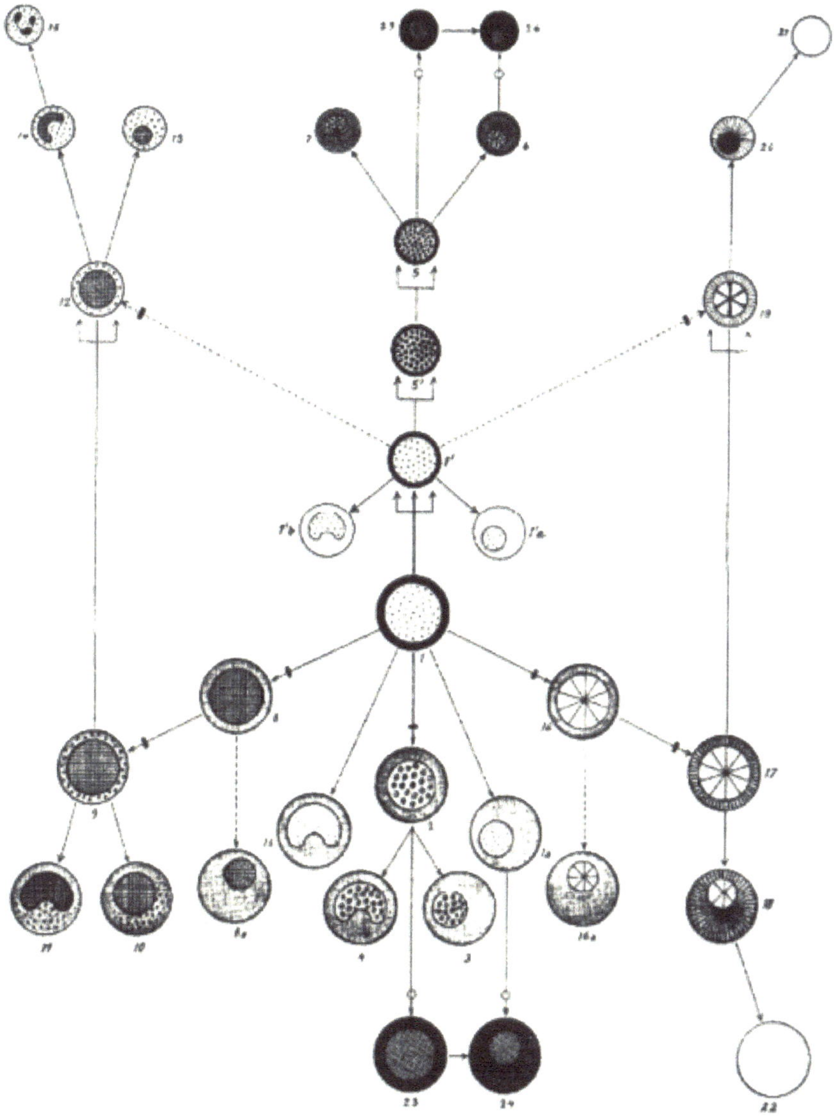

FIG. 1.2 Artur Pappenheim illustrated his view of hematopoiesis (1905). The cell in the center is the hypothesized common progenitor of all the blood cell lineages. Current known lineages of hematopoietic stem cells are illustrated in Fig. 1.4. *Reproduced with permission from Ramalho-Santos, M., Willenbring, H., 2007. On the origin of the term "stem cell." Cell Stem Cell 1, 35–38.*

the Rockefeller Institute. Dr. Florence Sabin was interested in the effects of X-ray use, which was becoming common at that time (Sabin et al., 1932). Sabin et al. discovered that the primary lymph node and bone marrow damage was primarily to chromatin in the nuclei of stem cells leading to eventual atrophy of these structures. Four years later Dr. Sabin along with a different team published a study on the development of blood and bone marrow in rabbits that includes a discussion on whether the stem cell for all the white blood cells was the same stem cell that gives rise to lymphocytes (Sabin, 1936). Important work in the 1950s on bone marrow transplants in dogs culminated in the 1957 paper that demonstrated the regenerative properties of bone marrow after lethal radiation, lending further support to the stem cell theory (Thomas, 1957). Dr. E. Donnall Thomas received the Nobel Prize in 1990 for demonstrating that after using radiation to kill bone marrow stem cells in a leukemia patient, an infusion of bone marrow from a healthy donor restored the patient to a leukemia-free life (Thompson et al., 1986).

Advancing Stem Cell Research

In 1953 James Watson and Francis Crick published the structure of DNA (Watson and Crick, 1953). The Nobel Prize-winning discovery was relevant to understanding that Weismann's germ plasm is DNA, genetic material that is replicated when cells divide. Research on gene expression and genetic regulation led to insights on the processes of self-renewal and differentiation. Stem cell studies that are based on our understanding of DNA have led to advances in tissue regeneration and disease modeling (Csobonyeiova et al., 2017; Rutkowski et al., 2017).

The discovery of stem cells is credited to Drs. James Till and Ernest McCulloch, honored for their research in 2005 with the Albert Lasker Award for Basic Medical Research. Till and McCulloch published a series of publications describing the regenerative and multipotent properties of hematopoietic stem cells. First, they determined the number of cells in a bone marrow transplant required to rescue mice that were irradiated with a lethal dose (Till and McCulloch, 1961). Next, they showed that cell colonies (nodules on the spleen) from the rescued animals arise from a single progenitor (Becker et al., 1963). Using an IBM 7090 they developed a model to account for self-renewal and differentiation in a heterogeneous colony of more than a million cells (Fig. 1.3) (Till et al., 1964). Further work with irradiated mice led to the observation that irradiated cells occasionally develop abnormal karyotypes, but can still divide. They used abnormal karyotype as a marker to determine that all the cells in a single colony are clones of the original (Becker et al., 1965). Next, they were able to concretely demonstrate that a single stem cell gives rise to all three types of mature blood cells—red cells, white cells and platelets (Fig. 1.4; Becker et al., 1965). Later they went on to investigate the factors that influence stem cell self-renewal and differentiation, opening up new avenues of investigation.

FIG. 1.3 An IBM 7090 used by Lasker Award winners Till and McCulloch to develop a model of self-renewal and differentiation in a heterogeneous colony of more than a million cells. *Image borrowed with permission from https://www-03.ibm.com/ibm/history/exhibits/mainframe/mainframe_PP7090.html.*

During this same period Drs. Joseph Altman and Gopal Das described cells in the dentate gyrus region of the hippocampus of rats that were labeled with a marker of dividing cells, tritiated thymidine (Altman and Das, 1965). Altman followed this study up with another finding in 1969 showing that the subventricular zone of the lateral ventricles contained cells that could also be labeled with tritiated thymidine (Altman, 1969). Unfortunately these data received no further attention for a decade and were much disbelieved until the late 1990s. The biggest hurdle to acceptance was that in 1928 Ramon y Cajal, considered the father of Neuroscience, proposed, "In the adult centers, the nerve pathways are fixed, and immutable: everything may die, nothing may be regenerated" (Ramon y Cajal, 1928). His words became the "no new neurons" dogma that still persists today outside the life sciences. We will discuss this exciting field of research in Section 2.

In 1968 Dr. Leroy Stevens, working at Jackson Laboratory, published a study on the isolation of fertilized 1-day (2-cell) mouse embryos and a subsequent testicular graft (Stevens, 1968). He had observed that the testes provided an environment that allowed the embryonic cell grafts to thrive. He also observed that some cells in the graft remained undifferentiated for more than 5 months (Stevens, 1968). In a follow-up study he examined

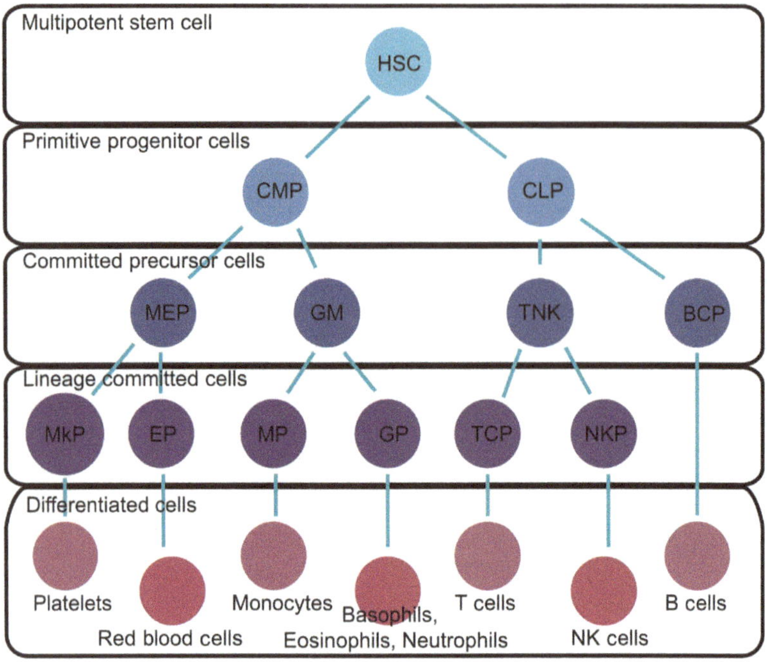

FIG. 1.4 Current known lineages of hematopoietic stem cells (HSC). HSCs give rise to intermediate progenitors that in turn give rise to stem cells with reduced potency. CMP, common myeloid progenitor; CLP, common lymphoid progenitor; MEP, megakaryocytes and erythroid cells; GM, granulocytes and macrophages; TNK, T cells and natural killer cells; BCP, B cell committed progenitor; MkP, megakaryocyte progenitor; EP, erythrocyte progenitor; MP, monocyte progenitor; GP, granulocyte progenitor; TCP, T cell progenitor; NKP, NK cell progenitor.

3- and 6-day embryos to observe the nature of the undifferentiated cells and *teratocarcinoma* growth from this later stage. Within the tumor he found cells could differentiate into *primordial germ cells*. He found that the tumors could be transplanted multiple times and attributed the nature of the tumor growth to "pluripotent embryonic stem cells" (Stevens, 1970). Stevens is acknowledged as the pioneer of embryonic stem cell (ESC) research. Stem cell lines were established from the teratocarcinomas called embryonic carcinoma cell lines and studied for their pluripotent properties.

In 1981 Dr. Gail Martin established a pluripotent stem cell line directly from ESCs she isolated from the inner cell mass of late blastocysts (Martin, 1981). An important aspect of the experiment was that the cells were cultured in a medium obtained from dishes in which an established teratocarcinoma stem cell line proliferated. This medium, called *conditioned media*, was thought to contain factors that either allowed the ESCs to proliferate or inhibited them from differentiating. That same year Drs. Martin Evans and Matthew Kaufman also published the isolation, culture and cultivation of mouse ESCs (Evans and Kaufman, 1981). Both studies contributed to the techniques involved in extracting the cells and determining the ideal culture conditions.

In 1998 Dr. James Thomson applied the accumulating knowledge of cell culture conditions for stem cells to human fertilized eggs. The eggs were obtained from couples undergoing in vitro fertilization (IVF) at a fertility clinic, and who had given their permission to use their leftover eggs for scientific research. First, Thomson demonstrated that the cells could proliferate in vitro for several months (Thomson et al., 1998). Second, by injecting the cells into immunocompromised mice, he showed that the cells could produce teratomas that contained differentiated cells representing all three embryonic *germ layers*. Thus Thomson showed that human ESCs have the potential to give rise to any cell in the body, suggesting a potential cure for myriad degenerative diseases.

CLONING

What do you think of when you hear the word "clone"? A TV show or movie? While the media has portrayed cloning as far-flung science fiction, people have been cloning organisms for more than 100 years! A clone can be defined as an organism with an identical genetic copy, like identical twins. The history of cloning began with the humble sea urchin, which develops from a single-celled embryo just like we do. The developmental process of sea urchins can be observed in a dish under a microscope as the embryo is transparent. Hans Adolf Edward Driesch performed the first documented cloning experiment using the sea urchin (Driesch, 1900, 1906). Driesch waited until the single-celled embryo divided into two cells; he then shook it until there were two single cells again (Fig. 1.5).

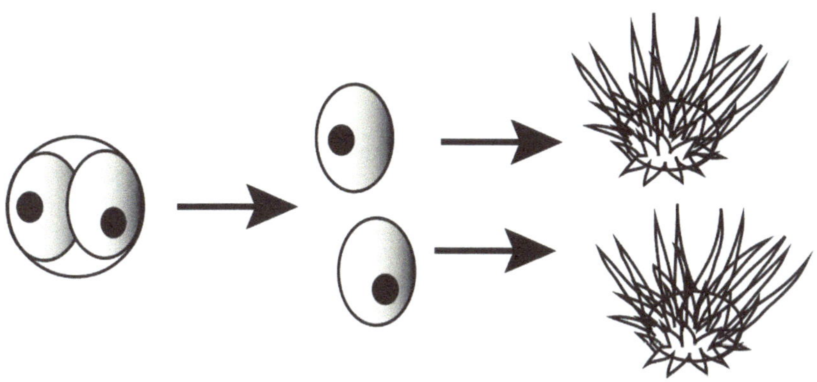

FIG. 1.5 The cloning of a sea urchin embryo. A single embryo at the two-cell stage was shaken in water until the cells separated. Each cell went on to produce a complete sea urchin.

Each cell went on to develop into a complete sea urchin, identical clones, demonstrating that early developing cells have the potential to become a full organism. Thus when a stem cell replicates its DNA and divides, a clone of the original stem cell is produced.

In 1995 Ian Wilmut at the Roslin Institute cloned a sheep named Dolly by using the DNA (nucleus) of a somatic cell and transferring it to an enucleated embryo (Campbell et al., 1996). The procedure, somatic cell nuclear transfer (SCNT), had been performed in frogs before, but Dolly was the first mammal created by nuclear transfer. Dolly developed from an embryo that had DNA identical to the somatic cell donor, which is what made her a clone. One could use SCNT to clone their cells and provide a way to test drugs on the cultured cells, or differentiate the cells to produce replacement tissue. The risk of host rejection would be eliminated. The drawback is that SCNT still requires the destruction of an embryo when its nucleus is removed.

The first nonhuman primate was cloned in 1997 and at this time, then US President Bill Clinton proposed a 5-year moratorium on cloning to provide time for scientists and other concerned parties to debate the possibility of cloning humans (Meng et al., 1997). However, in other countries or in the private sector cloning continued over the next 10 years using cows, mice, goats, gaur and mouflon, the last two belonging to endangered species (Lanza et al., 2000; Yazawa et al., 1997). Researchers have cloned both human and primate ESCs, creating stem cell lines that are used to study therapeutic applications (Thomson et al., 1995, 1998) (Box 1.1).

Ethics of ESCs and Cloning

The use of ESCs in research is a hotly debated topic. Some scientists proclaim the benefits of a pluripotent cell for research. ESCs have the potential to cure neurodegenerative diseases, genetic disorders, injuries, and even create new organs for those requiring a transplant. An argument against ESC is that their use in medicine is accompanied with the risk of allogenic rejection. Along these lines, some argue that the cost for individualized treatment using ESC is unattainable for most and may require lifetime immunosuppressant treatment. The cost of individual treatment invokes further ethical issues concerning justice and health care. Equal access and the fair distribution of healthcare resources is another related debate. Some opposed to ESC research offer umbilical cord blood (UCB) as an alternative. It was recently shown that stem cells from UCB can differentiate into other cell types such as neurons (Ali et al., 2013). UCB will have to undergo the same exacting study that has been done for ESCs to determine whether UCB is an acceptable replacement in any situation in which one would use ESCs.

BOX 1.1

SPOTLIGHT ON ADVANCED CELL TECHNOLOGY

Advanced Cell Technology (ACT) is a company located in Worcester, MA that started out working on agricultural cloning research. ACT received a great deal of attention when it announced in 2001 the successful cloning of human embryos from adult cells in order to derive stem cells compatible with the donor (Lanza et al., 2001). Vice president and tissue engineering scientist Robert Lanza published a proof of principle paper in 2006 demonstrating the ability to derive human ESC lines from blastomeres, cells that make up the early embryo (Klimanskaya et al., 2006).

IVF clinics must harvest multiple eggs from a woman that are screened after fertilization to determine the best candidate(s) for implantation. With permission, ACT used embryos from a clinic that were not selected for implantation. The embryos were cultured to the 8–10 cell stage at which point 4–7 blastomeres (cells) were removed using a biopsy procedure (Klimanskaya et al., 2006). Single blastomere removal is a technique used widely by IVF clinics for preimplantation genetics to screen for diseases or genetic abnormalities, without harming the embryo. Lanza's group cultured blastomeres from 10 embryos from which they obtained two stable lines of hESCs that proliferated without any sign of differentiation for 8 months (Klimanskaya et al., 2006). However, the removal of 4–7 cells resulted in the destruction of the embryo. A subsequent paper (Chung et al., 2008) repeated the experiments using enhanced culture media and removed only a single blastomere (Chung et al., 2008). The study was again able to establish four hESC lines but avoided the ethical issue of destroying an embryo.

ACT has focused their stem cell work on ophthalmological problems and in 2011 applied for a clinical trial to test the transplantation of hESC-derived retinal cells to treat Stargardt's macular dystrophy, a genetic disease that causes vision loss due to degeneration of the macula. In the United States approximately 1%–2% of people over 50 years old and up to 14% of Caucasians over 80 years old are affected ([NEI] National Eye Institute, 2015). They have now completed Phase 2 clinical trials on both Stargardt's macular dystrophy and age-related macular degeneration (Schwartz et al., 2015). Long-term follow-up studies are currently underway. This is exciting news for those who suffer from these diseases, for which there are currently no treatments or cures.

Other commonly voiced opinions for cloning include producing desirable traits in animals, either for animal enthusiasts or livestock production. On one hand, the decrease in genetic diversity may have unexpected consequences. On the other hand, if genetic samples are available there is the potential to clone extinct or endangered species. Supporters of the deextinction debate argue that if humans are responsible for the extinction of a species then it is our responsibility to resurrect the species (Shultz, 2017). On the other side of the debate, the cost of investing in the development and care of a once-extinct species could go toward preserving current near-extinct species. Cloning for these purposes may seem ethically acceptable (though not completely without problems), but the biggest debate is on issues regarding humans.

The previous cloning descriptions all required an embryo. In opposition, the destruction of an embryo is unacceptable to those who consider fertilization of the egg the beginning of life. Therefore does the embryo have the same right to life as a human fetus, as a child or adult? Depending on the moral status of an embryo, the legal status differs. A bunch of cells may not have legal rights, but does an embryo have legal rights? Some argue that the value of an embryo is greater than the treatment of a disease. That argument has led to a discussion of comparing value: the potential of an embryo or the life already being lived. Is it worth sacrificing one embryo for the sake of the many? Some may find the idea of cloning a person who has died offensive, while others would jump at the opportunity to have their loved one with them again. Other arguments against cloning include physical (birth defects)/mental/emotional problems for the cloned human. Further, some people oppose cloning for religious or moral reasons, while others argue that scientists should not take nature into their own hands. There is no simple solution to these debates, however, research continues to explore the power of stem cells and cloning under current federal guidelines or with private funds. One possibility is that scientists will find a way to capitalize the potential of stem cells without harming an embryo.

An important stem cell discovery that may alter the ethics debate is *induced* pluripotent stem cells (iPSCs). These are somatic cells that have been reprogrammed to express specific genes causing them to revert from a differentiated state to a pluripotent state. The discovery was the result of many research groups working to determine what gave a cell "stemness." Shinya Yamanaka discovered four genes that when expressed together could turn *fibroblasts* into iPSCs (Takahashi and Yamanaka, 2006). This recent discovery has led to new avenues of investigation, such as the induction of different somatic cell types into stem cells, comparisons of iPSCs to ESCs, and the potential of iPSCs for treating disease. Yamanaka's original 2006 article has been cited more than 10,000 times in subsequent research papers, demonstrating its huge impact on stem cell science.

POLITICAL HISTORY OF STEM CELL RESEARCH REGULATION

The following list details some of the decisions made by the judicial, legislative or executive branches of the US government that have either directly or indirectly affected stem cell research. The decisions initially affected access to ESCs due to abortion or IVF treatment. Later decisions directly banned or allowed research using ESCs.

- 1973 Supreme Court decision Roe v. Wade legalizes abortion during the first two trimesters
- 1974 Congress halts federal funding of research on embryos or embryonic tissue
- 1978 The first "test tube" baby is born to the first parents to successfully undergo IVF
- 1987 Fetal tissue research banned when researchers ask to transplant fetal cells in Parkinson's Disease patients
- 1993 President Clinton lifts ban allowing research on donated IVF embryos and aborted fetal tissue
- 1995 President Clinton reverses his lift of the ban except on *existing* human ESC lines
- 1997 President Clinton bans federal funds for human cloning research
- 2001 President Bush allows federal funding of ESC research exclusively on 60 *existing* stem cell lines that were created from embryos left over from IVF
- 2002 California Governor Gray Davis signs a law that allows research on stem cells from fetal and embryonic tissue, including cloning embryos using SCNT for the purpose of treating disease
- 2004 A California ballot initiative passes approving the creation of the California Institute for Regenerative Medicine, making stem cell research a constitutional right in that state
- 2009 President Obama removes the restriction on federal funding for newer stem cell lines, allowing researchers to use cell lines created in the private sector
- 2016 States that allow stem cell research or have stem cell institutions (34): Alabama, California, Colorado, Connecticut, Delaware, Florida, Georgia, Hawaii, Illinois, Indiana, Iowa, Kansas, Kentucky, Maine, Maryland, Massachusetts, Michigan, Minnesota, Missouri, Nebraska, Nevada, New Hampshire, New Jersey, New York, North Carolina, Ohio, Oklahoma, Oregon, Pennsylvania, Rhode Island, Texas, Utah, Washington, and Wisconsin

SUMMARY

Work from several fields of research came together to produce what we know about stem cells including early work in genetics, cancer and IVF. The promise of ESCs captured the imagination of scientists who would use them to find ways to cure diseases. The ability of a cell to self-renew or become any cell type provides the potential to replace any cell in the body. However, the ethical issues of obtaining ESCs from human embryos have slowed progress on one hand, while spurring innovative research in iPSCs on the other hand. Current federal and state laws on ESC lines allow limited research, but if the list of political decisions is any indication, the allowances are subject to change. Advances in iPSC research may make access to ESCs less of an issue, though an understanding of these cells is essential to fields such as development, cancer and degeneration. The continued ability of scientists to study ESCs has the potential for major discoveries, applications and cures.

DEFINITIONS

Conditioned media Media taken from a dish with a specific cell type growing in it. The cells release undefined biologically active factors.

Differentiate A stem or progenitor cell divides and one or both daughter cells become a more specialized cell type.

Fibroblasts The most common connective tissue cell in animals that produces extracellular matrix and collagen.

Germ cell Sperm or ovum containing one set of chromosomes after meiosis that can combine with a germ cell from the opposite sex resulting in an embryo.

Germ layers Early cell specifications of the embryo that give rise to explicit tissue types in the body.

Mesenchymal cells Arising from the germ layer called the mesenchyme.

Pluripotent The potential to become any cell type in the body.

Primordial germ cells The cells of the embryo that develop into stem cells that produce germ cells, sperm or ovum.

Self-renewal When stem cells divide to produce one or two new copies of the original.

Somatic cell Adult or terminally differentiated cells of the body containing a full set of chromosomes. Not a reproductive cell.

Teratocarcinoma A combination of two terms, teratoma and embryonic carcinoma. It is a malignant germ cell growth in the testes in which differentiated cells from all three germ layers can be found. In vivo it has the potential to metastasize and invade local tissues. However, ESC-derived masses are typically benign (Lensch and Ince, 2007).

References

Ali, H., Bayatti, N., Lindsay, S., Dashti, A.A., Al-Mulla, F., 2013. Directed differentiation of umbilical cord blood stem cells into cortical GABAergic neurons. Acta Neurobiol. Exp. (Warsz.) 73, 250–259.

Altman, J., 1969. Autoradiographic and histological studies of postnatal neurogenesis. IV. Cell proliferation and migration in the anterior forebrain, with special reference to persisting neurogenesis in the olfactory bulb. J. Comp. Neurol. 137, 433–457.
Altman, J., Das, G.D., 1965. Autoradiographic and histological evidence of postnatal hippocampal neurogenesis in rats. J. Comp. Neurol. 124, 319–335.
Becker, A.J., McCulloch, E.A., Till, J.E., 1963. Cytological demonstration of the clonal nature of spleen colonies derived from transplanted mouse marrow cells. Nature 197, 452–454.
Becker, A.J., Mcculloch, E.A., Siminovitch, L., Till, J.E., 1965. The effect of differing demands for blood cell production on DNA synthesis by hemopoietic colony-forming cells of mice. Blood 26, 296–308.
Boveri, T., 1889. Ein geschlechtlich erzeugter Organismus ohne mü tterliche Eigenschaften [An organism sexually produced without characteristics of the mother]. Sitz. Gesel. Morph. Physiol. Müchen. 5, 73–83.
Campbell, K.H.S., McWhir, J., Ritchie, W.A., Wilmut, I., 1996. Sheep cloned by nuclear transfer from a cultured cell line. Nature 380, 64–66.
Chen, M., Przyborowski, M., Berthiaume, F., 2009. Stem cells for skin tissue engineering and wound healing. Crit. Rev. Biomed. Eng. 37, 399–421.
Chung, Y., Klimanskaya, I., Becker, S., Li, T., Maserati, M., Lu, S.-J., Zdravkovic, T., Ilic, D., Genbacev, O., Fisher, S., et al., 2008. Human embryonic stem cell lines generated without embryo destruction. Cell Stem Cell 2, 113–117.
Csobonyeiova, M., Polak, S., Zamborsky, R., Danisovic, L., 2017. iPS cell technologies and their prospect for bone regeneration and disease modeling: a mini review. J. Adv. Res. 8, 321–327.
Danchakoff, V., 1916. The differentiation of cells as a criterion for cell identification, considered in relation to the small cortical cells of the thymus. J. Exp. Med. 24, 87–105.
Darwin, C., 1859. On the Origin of Species (by Means of Natural Selection).
Driesch, H., 1900. Die isolierten Blastomeren des Echinodenkeimes. Wilhelm Roux's Arch. Dev. Biol 10, 361–410.
Driesch, H., 1906. Studien zur Entwicklungsphysiologie der Bilateralitat. Wilhelm Roux's Arch. Dev. Biol 21, 756–791.
Evans, M.J., Kaufman, M.H., 1981. Establishment in culture of pluripotential cells from mouse embryos. Nature 292, 154–156.
Häcker, V. (1892). Zool. Jahrbucher, Jena.
Haeckel, E., 1905. Anthropogeny: Or, the Evolutionary History of Man, fifth ed. Wilhelm Engelmann.
Klimanskaya, I., Chung, Y., Becker, S., Lu, S.-J., Lanza, R., 2006. Human embryonic stem cell lines derived from single blastomeres. Nature 444, 481–485.
Lanza, R.P., Cibelli, J.B., Diaz, F., Moraes, C.T., Farin, P.W., Farin, C.E., Hammer, C.J., West, M.D., Damiani, P., 2000. Cloning of an endangered species (*Bos gaurus*) using interspecies nuclear transfer. Cloning 2, 79–90.
Lanza, R.P., Cibelli, J.B., West, M.D., Dorff, E., Tauer, C., Green, R.M., 2001. The ethical reasons for stem cell research. Science 292, 1299.
Lensch, M.W., Ince, T.A., 2007. The terminology of teratocarcinomas and teratomas. Nat. Biotechnol. 25 (11), 1211–1212.
Martin, G.R., 1981. Isolation of a pluripotent cell line from early mouse embryos cultured in medium conditioned by teratocarcinoma stem cells. Proc. Natl. Acad. Sci. U. S. A. 78, 7634–7638.
Maximow, A., 1909. Fol. Haematol 8, 125–134.
Maximow, A., 1924. Relation of blood cells to connective tissues and endothelium. Physiol. Rev. 4, 533–563.
Meng, L., Ely, J.J., Stouffer, R.L., Wolf, D.P., 1997. Rhesus monkeys produced by nuclear transfer. Biol. Reprod. 57, 454–459.

National Eye Institute, 2015. Facts About Stargardt Disease.
Ramalho-Santos, M., Willenbring, H., 2007. On the origin of the term "stem cell". Cell Stem Cell 1, 35–38.
Ramon y Cajal, S., 1928. Degeneration and Regeneration of the Nervous System. Haffner Publishing Co., New York, NY.
Rutkowski, T.P., Schroeder, J.P., Gafford, G.M., Warren, S.T., Weinshenker, D., Caspary, T., Mulle, J.G., 2017. Unraveling the genetic architecture of copy number variants associated with schizophrenia and other neuropsychiatric disorders. J. Neurosci. Res. 95, 1144–1160.
Sabin, F.R., Doan, C.A., Forkner, C.E., 1932. The production of osteogenic sarcomata and the effects on lymph nodes and bone marrow of intravenous injections of radium chloride and mesothorium in rabbits. J. Exp. Med. 56 (2), 267–289.
Sabin, F., 1936. Changes in the bone marrow and blood cells of developing rabbits. J. Exp. Med. 64 (1), 97–120.
Schwartz, S.D., Regillo, C.D., Lam, B.L., Eliott, D., Rosenfeld, P.J., Gregori, N.Z., Hubschman, J.-P., Davis, J.L., Heilwell, G., Spirn, M., et al., 2015. Human embryonic stem cell-derived retinal pigment epithelium in patients with age-related macular degeneration and Stargardt's macular dystrophy: follow-up of two open-label phase 1/2 studies. Lancet Lond. Engl. 385, 509–516.
Shultz, D., 2017. Bringing extinct species back from the dead could hurt—not help—conservation efforts. Science. Accessed on December 22, 2018, Available at: http://www.sciencemag.org/news/2017/02/bringing-extinct-species-back-dead-could-hurt-not-help-conservation-efforts.
Stevens, L.C., 1968. The development of teratomas from intratesticular grafts of tubal mouse eggs. J. Embryol. Exp. Morphol. 20, 329–341.
Stevens, L.C., 1970. The development of transplantable teratocarcinomas from intratesticular grafts of pre- and postimplantation mouse embryos. Dev. Biol. 21, 364–382.
Takahashi, K., Yamanaka, S., 2006. Induction of pluripotent stem cells from mouse embryonic and adult fibroblast cultures by defined factors. Cell 126, 663–676.
Thomas, E., 1957. Intravenous infusion of bone marrow in patients receiving radiation and chemotherapy. N. Engl. J. Med. 257, 491–496.
Thompson, C.B., Sanders, J.E., Flournoy, N., Buckner, C.D., Thomas, E.D., 1986. The risks of central nervous system relapse and leukoencephalopathy in patients receiving marrow transplants for acute leukemia. Blood 67, 195–199.
Thomson, J.A., Kalishman, J., Golos, T.G., Durning, M., Harris, C.P., Becker, R.A., Hearn, J.P., 1995. Isolation of a primate embryonic stem cell line. Proc. Natl. Acad. Sci. U. S. A. 92, 7844–7848.
Thomson, J.A., Itskovitz-Eldor, J., Shapiro, S.S., Waknitz, M.A., Swiergiel, J.J., Marshall, V.S., Jones, J.M., 1998. Embryonic stem cell lines derived from human blastocysts. Science 282, 1145–1147.
Till, J.E., McCulloch, E.A., 1961. A direct measurement of the radiation sensitivity of normal mouse bone marrow cells. Radiat. Res. 14, 213–222.
Till, J.E., Mcculloch, E.A., Siminovitch, L., 1964. A stochastic model of stem cell proliferation, based on the growth of spleen colony-forming cells. Proc. Natl. Acad. Sci. U. S. A. 51, 29–36.
Watson, J.D., Crick, F.H., 1953. The structure of DNA. Cold Spring Harb. Symp. Quant. Biol. 18, 123–131.
Weismann, A., 1892. Das Keimplasma: Eine Theorie der Vererbung. Gustav Fischer, Jena.
Yazawa, S., Aoyagi, Y., Konishi, M., Takedomi, T., 1997. Characterization and cytogenetic analysis of Japanese Black calves produced by nuclear transfer. Theriogenology 48, 641–650.

CHAPTER

2

Stem Cells: From Embryos to Adults

OUTLINE

LEARNING OBJECTIVES

1. Explain the difference between symmetric and asymmetric stem cell division.
2. Describe what influences the modes of proliferation.
3. Discuss the origin of the three germ layers and describe the tissue/organs and somatic stem cells that arise from each layer.

https://doi.org/10.1016/B978-0-12-811014-0.00002-0

4. Describe what may influence or regulate stem cell proliferation.
5. Describe the advantages and disadvantages of embryonic and somatic stem cells.

INTRODUCTION

Upon fertilization the zygote has the potential to differentiate or to become any cell in the body. Over the course of development the majority of cells in the organism differentiate to a final fate. However, a small number of cells, the stem cells, remain after birth and retain varying levels of potential to become other cells. A stem cell can divide to renew itself in a controlled way over the entire lifespan of an organism. This chapter introduces some of the different types of stem cells found during embryogenesis and identifies those that remain postnatally and their associated potencies.

A major goal in stem cell research is to harness stem cell potential in order to produce a desired cell type. Understanding the mechanisms regulating proliferation is essential to accomplish this goal. The Delta/Notch signaling pathway is one mechanism that influences proliferation in embryonic and postnatal cell division. We will review how this pathway regulates the fate of proliferating cells. Discovering the mechanisms affecting proliferation is not only important for stem cell research but also has implications for understanding development, regeneration and cancer.

STEM CELL PROPERTIES

When stem cell division by mitosis produces two new identical stem cells it is called *symmetric cell division* or *self-renewal* (Fig. 2.1, top); self-renewal is one of the defining features of stem cells (Morrison and Kimble, 2006). Symmetric cell division occurs if the contents of a cell, including the chromosomes, are divided equally during mitosis, resulting in daughter cells that have the same potency as the original, mother cell. Potency is the second defining feature of stem cells and refers to the potential of stem cells to produce more than one type of cell. *Asymmetric cell division* allows a stem cell to produce one cell identical to itself (self-renewal) and one cell that is different from the mother cell (Fig. 2.1, bottom). The new, unique daughter cell may only be able to differentiate into a specific cell type, or it may be multipotent, but unable to give rise to as many cell types as the mother cell (Lin and Schagat, 1997).

During cell division the potency of a cell is determined due to regulation by internal factors such as orientation of the mitotic spindle (Kaltschmidt et al., 2000), and external factors such as where a cell resides in relation to other cells or tissues. During mitosis the chromosomes are duplicated and then arranged across the middle of the cell. The mitotic spindle, which consists of two centromeres present at opposite ends of the cell, extends microtubules that attach to the chromosomes. The duplicated pairs are

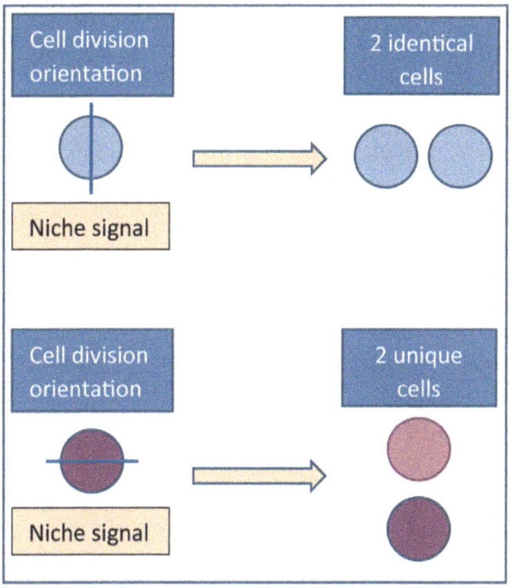

FIG. 2.1 The influence of the niche and mitotic spindle orientation on cell division. Top, each daughter cell is equally exposed to the niche when the orientation of the mitotic spindle is perpendicular to the niche. The two daughter cells are identical. Bottom, one daughter cell is exposed to niche signals while the other is not when the orientation of the mitotic spindle is parallel to the niche. In this manner cell division produces two unique daughter cells.

pulled apart through a shortening process of the microtubules, bringing one chromosome copy to each side of the cell. Finally, the cell undergoes cleavage in which division and separation of the cytoplasm results in two new cells. When the mitotic spindle aligns perpendicular to the apical/basal plane of the cell relative to the environment fate determinants are divided equally, giving rise to identical daughter cells (Yamashita et al., 2007). Asymmetric cell division occurs when the mitotic spindle aligns parallel to a regulating determinant. For example, in *Drosophila*, the mitotic spindle of sensory organ precursor cells can align parallel to the basal plane of the cell resulting in the unequal distribution of internal fate determinants between the two daughter cells (Knoblich, 2008). One daughter cell is a self-renewal of the stem cell, while the other expresses different genes, altering its cell fate. Thus although the two daughter cells each receive a copy of the chromosomes, they are not identical after asymmetric cell division.

Another factor that can influence a stem cell division is its location. Stem cells often reside in a special environment called the stem cell *niche* (Chapter 3). The niche provides external factors that can also influence how a stem cell will divide. When the plane of division is parallel to the niche, the cell in contact with the niche may take on a fate different from the cell that is produced separated from the niche (Fig. 2.1, bottom; Yamashita, 2003). The niche may influence stem cell potency, either maintaining potency (self-renewal) or inducing a more committed fate.

A fertilized egg, or zygote, has the potential to produce any type of cell in the body. An embryo is *totipotent*: it has the potential to produce every cell in the body *and* the placenta (De Paepe et al., 2014). Thus, totipotency is only a brief period as the embryo begins to divide and differentiate. The fertilized egg (day one) undergoes mitosis followed by cleavage division to form a two-cell embryo with each cell repeating this process until the 16-cell stage is reached (Rossant and Tam, 2009). At this stage the cells called blastomeres form tight junctions with one another and continue dividing. By day four, the cells have organized themselves to form a hollow, fluid-filled ball of cells called the blastocyst (Rossant and Tam, 2009). The cells of the blastocyst are no longer totipotent. The outer blastocyst cells are now part of what will become the placenta. Within the blastocyst is a small cluster of cells called the inner cell mass; these are pluripotent *embryonic stem cells (ESCs)*. ESCs are the only cells of the blastocyst with the potential to become any cell in the body (Saiz and Plusa, 2013).

Following development of the blastocyst stage the embryo undergoes gastrulation. Gastrulation is a phenomenon in which three germ layers are produced that will give rise to specific tissue types (Fig. 2.2). The three germ layers are the endoderm, mesoderm and ectoderm, terms meaning inner, middle and outer, respectively. The endoderm gives rise to internal organs such as gut, liver and lungs. The mesoderm gives rise to muscles, vasculature and skeleton, and the ectoderm gives rise to the epidermis and nervous system. Germ layer cells are considered *multipotent*; they have lost the potential to produce an entire embryo but are capable of producing all cell types *within* each germ layer.

Certain multipotent cells remain in the body throughout life called *somatic stem cells* (Table 2.1). For example, as we learned in Chapter 1, hematopoietic stem cells are present throughout life, and give rise to all red and white blood (immune system) cells (Abramson et al., 1977). Other stem cells differentiate further and become unipotent, meaning they can only produce one cell type or tissue. For example, germline stem cells only produce gametes; epidermal basal cells only generate keratinocytes; satellite cells only produce new muscle (Kimble, 2014; Visvader and Clevers, 2016).

During development there may be times when a stem cell switches between symmetric or asymmetric cell division (Kaltschmidt et al., 2000). At the beginning of any particular development stage it is important to ensure that there are enough stem cells to produce a particular tissue type. Thus early in development *proliferative symmetric* cell division is favored to produce enough stem cells. Stem cells can then switch to *generative asymmetric* cell division to generate one new stem cell and one specialized cell. At the end of development stem cells may switch to a *generative symmetric* cell division in which a stem cell divides to become two committed cells, thus shrinking the stem cell population (Huttner and Kosodo, 2005; Morrison and Kimble, 2006).

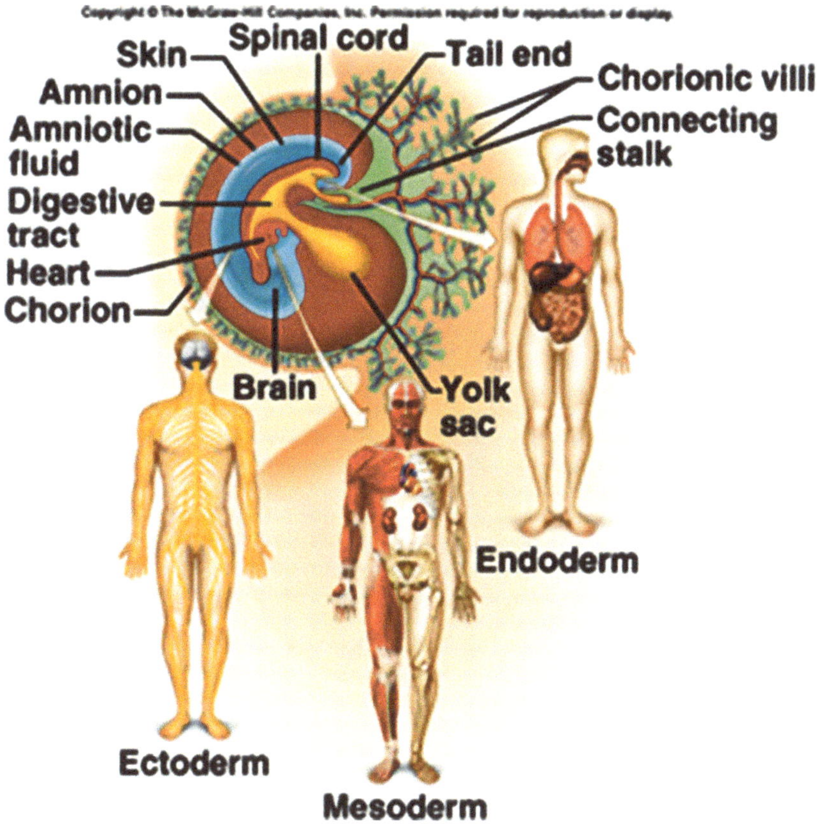

FIG. 2.2 The embryo undergoes gastrulation during which three germ layers are formed that will each give rise to specific tissue types.

HARNESSING STEM CELLS

Isolating ESCs

ESCs isolated from the inner cell mass of the blastocyst stage can be grown in culture and maintained as pluripotent cells or allowed to differentiate into a desired cell type (Fig. 2.3). Human ESCs can be harvested from a fertilized embryo that has been kept in optimal culture media to ensure normal growth. At the four- to five-day blastocyst stage, the inner cell mass consists of approximately 30–50 cells. Using enzymatic proteins to lyse the outer layer of the blastocyst, called the trophectoderm, the cells of the inner cell mass are separated and cultured in a dish that may contain a layer of mouse embryonic fibroblasts or specialized culture media (Gardner and Lane, 1998; Thomson et al., 1995). Self-renewal allows a stem cell to divide continuously, each time replacing itself with two

TABLE 2.1 A List of Some of the Known Somatic Stem Cell Types, Their Location in the Body and the Cells to Which They Give Rise

Type	Location	Produces
Endodermal origin		
Biliary tree	Liver	Pancreatic islet-like cells
Gut	Gut	Intestinal cells
Airway basal epithelia	Lungs	Lung epithelial cells
Mesodermal origin		
Umbilical cord blood	Umbilical cord	Pluripotent
Amniotic fluid	Uterus	Pluripotent
Hematopoietic	Peripheral blood Red bone marrow	Red blood cells, white blood cells (Immune cells), blood vessels
Hair follicle	Dermal papillae	Hair
Cardiac stem cells	Heart	Myocytes
Satellite cells	Skeletal muscle	Muscle
Osteoblast	Bone marrow	Bone and cartilage (chondrocytes)
GLI1+ progenitors	Adrenal capsule	Gonadal-like tissue
Fat/adipocytes	Adipose tissue	Adipocytes
Ectodermal origin		
Dental	Teeth	Dental pulp, dentin
Neural	Spinal cord, forebrain dentate gyrus	Neurons, glia
Skin	Skin	Skin

Carraro, G., Perin, L., Sedrakyan, S., Giuliani, S., Tiozzo, C., Lee, J., Turcatel, G., De Langhe, S.P., Driscoll, B., Bellusci, S., et al., 2008. Human amniotic fluid stem cells can integrate and differentiate into epithelial lung lineages. Stem Cells Dayt. Ohio 26, 2902–2911; De Coppi, P., Bartsch, G., Siddiqui, M.M., Xu, T., Santos, C.C., Perin, L., Mostoslavsky, G., Serre, A.C., Snyder, E.Y., Yoo, J.J., et al., 2007b. Isolation of amniotic stem cell lines with potential for therapy. Nat. Biotechnol. 25, 100–106; Dörner, J., Martinez Rodriguez, V., Ziegler, R., Röhrig, T., Cochran, R.S., Götz, R.M., Levin, M.D., Pihlajoki, M., Heikinheimo, M., Wilson, D.B., 2017. GLI1 + progenitor cells in the adrenal capsule of the adult mouse give rise to heterotopic gonadal-like tissue. Mol. Cell. Endocrinol. 441, 164–175; Kolambkar, Y.M., Peister, A., Soker, S., Atala, A., Guldberg, R.E., 2007. Chondrogenic differentiation of amniotic fluid-derived stem cells. J. Mol. Histol. 38, 405–413; Perin, L., Giuliani, S., Jin, D., Sedrakyan, S., Carraro, G., Habibian, R., Warburton, D., Atala, A., De Filippo, R.E., 2007. Renal differentiation of amniotic fluid stem cells. Cell Prolif. 40, 936–948; Sun, H., Feng, K., Hu, J., Soker, S., Atala, A., Ma, P.X., 2010. Osteogenic differentiation of human amniotic fluid-derived stem cells induced by bone morphogenetic protein-7 and enhanced by nanofibrous scaffolds. Biomaterials 31, 1133–1139.

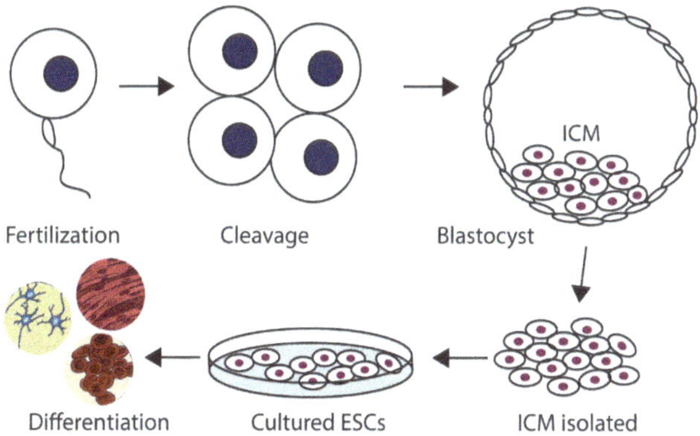

FIG. 2.3 Isolation of embryonic stem cells from a fertilized embryo at the blastocyst stage. The inner cell mass (ICM) consists of pluripotent stem cells that can be cultured in vitro and differentiated into other cell types.

identical cells. Cells cultured this way will continue to divide symmetrically up to at least 5 months (Thomson et al., 1998).

In Vivo Growth

In order to determine how cultured ESCs behave in vivo, scientists injected them into mice with compromised immune systems (to avoid immune rejection). Depending on where the cells were injected, formation of tumors was commonly observed consisting of multiple cell types, such as intestinal cells, neural epithelium, bone, cartilage, and striated muscle (Stevens, 1970). The cell types observed in the tumors are representatives of the three germ layers, demonstrating the pluripotency of the cells. These experiments also reveal a risk of using ESCs—uncontrolled cancerous growth.

In Vitro Growth

Scientists have discovered ways to control the differentiation of ESCs to achieve a desired cell type in a dish (Fig. 2.3). Altering the culture media with growth hormones or other determinants will produce differentiated cell types. For example, reducing the expression of a specific proteoglycan in mouse ESCs increased the dopaminergic neuron differentiation, the neuron type that degenerates in Parkinson disease patients (Fico et al., 2014). ESCs have been cultured to produce neural precursors, hematopoietic stem cells, and cardiac myocytes (He et al., 2003; Kaufman et al., 2001; Zhang et al., 2001). ESCs have the potential for major medical progress, but because of the risk of tumor development the field has progressed with slow and careful steps.

ESC-LIKE SOURCES

Umbilical Cord Blood Stem Cells

In 2002, Colin P. McGuckin discovered a new source of ESC-like cells in umbilical cord blood that can be harvested without harming an embryo (Forraz et al., 2002). Further experiments showed that these cells expressed markers similar to ESCs and could be differentiated into both hematopoietic and neuroglial progenitors (McGuckin et al., 2004). Since then, many researchers have focused on the pluripotent potential of umbilical cord blood cells (Harris and Rogers, 2007). These ESC-like cells have been induced to differentiate into insulin-producing cells (β cells), neural cells and hepatocytes (liver cells) (Denner et al., 2007; McGuckin et al., 2008; Moon et al., 2008). Another advantage of these cells is that they expand in culture easily without the use of fibroblasts or a feeder layer. The feeder layer is often obtained from nonhuman sources such as mice. Therefore, the risk of ESCs with possible contamination of pathogens harmful to humans is greatly decreased (Klimanskaya et al., 2005).

Amniotic Fluid Stem Cells

In 2003 scientists isolated another type of pluripotent stem cell from human amniotic fluid (In 't Anker et al., 2003). Human *amniotic fluid stem cells* (AFSCs) have a level of potency somewhat less than the pluripotency of ESCs but more than the traditional definition of a multipotent stem cell (Fauza, 2004; Witkowska-Zimny and Wrobel, 2011). Several characteristics make AFSCs a good alternative to blastocyst-derived ESCs (Abdulrazzak et al., 2013). First, it is possible to harvest the cells without harm to the mother or the fetus. Second, the cells are able to grow rapidly in culture without a feeder layer, doubling in number in approximately 36 h. Third, the cells do not show signs of tumorigenicity when injected into immune-compromised mice. Finally, AFSCs can produce cells of all three germ layers; researchers have induced AFSCs to differentiate into multiple cell types including renal, lung epithelial, chondrogenic (cartilage), smooth muscle, myocardial tissue and osteoblastic cells (Carraro et al., 2008; De Coppi et al., 2007a, 2007b; Kolambkar et al., 2007; Perin et al., 2007; Sun et al., 2010). Work by Bonaventura et al. comparing the potential of stem cells isolated from bone marrow, umbilical cord blood, human endometrium and amniotic fluid to differentiate into neuronal cell types found that AFSCs responded best to the neurodifferentiation media, expressing neural markers and electrophysiological properties (Bonaventura et al., 2015). Thus, AFSCs show promise for treating neurological injuries and diseases such as stroke and Huntington, Parkinson and Alzheimer diseases because of the ability to differentiate into neurons. The accessibility and potency of AFSCs make them one of the most exciting new sources of stem cells today.

SOMATIC STEM CELLS

As the development of each organ system is completed, a small population of stem cells remains, even after birth, and can be found throughout life. These remaining stem cells are called somatic or adult stem cells. Some somatic stem cells are multipotent and can give rise to several cell types. Other somatic cells are *unipotent*, only giving rise to a single cell type in addition to its own self-renewal. Somatic stem cells reside within tissues or organs, typically in a specialized niche of differentiated cells, for the purpose of healing, repairing, or replenishing the tissue(s) in which it resides in response to injury or daily wear and tear.

Somatic stem cells originate after gastrulation and therefore can have an endodermal, mesenchymal or ectodermal origin. Thus somatic stem cells found in the skin arise from the ectoderm; satellite cells in skeletal muscle arise from the mesoderm; intestinal stem cells in the gut arise from the endoderm. Hematopoietic stem cells described in Chapter 1 are mesenchymal somatic stem cells that repair bone and replenish the blood supply. Somatic stem cells found in soft tissues such as the liver and intestines have an endodermal origin. The nervous system has an ectodermal origin, and somatic stem cells have been discovered in two regions of the brain, the subventricular zone of the lateral ventricles and the dentate gyrus of the hippocampus. The neural stem cells in these two areas are the focus of Section 2 of this book. Other known locations of somatic stem cells are teeth, blood vessels, ovarian epithelium and testis.

When isolated from the body, somatic stem cells have shown a limited capacity to divide, making in vitro studies difficult to conduct. In order to demonstrate that a population of cells has somatic stem cell characteristics, a study will typically use two different tests. The first test uses research animals and injects a marker that labels proliferating cells (Box 2.1; Altman and Das, 1965). The animal is sacrificed and the tissue of interest is examined to determine which cells were proliferating. Another method is to remove a population of somatic stem cells from a living animal and infect them in culture with a genetic marker such as green fluorescent protein (GFP). The infected cells are then transplanted back into another animal of the same species and observed for the ability to replace the tissue from which they were taken, indicated by an increasing number of GFP+ cells (Tamplin et al., 2015). While these methods have improved the ability to identify somatic cells, questions regarding their potency, proliferation and regulation remain.

BOX 2.1

PROLIFERATION MARKERS

There are three ways in which cell proliferation can be observed: the labeling of DNA synthesis, cellular metabolism or proliferation proteins. Each method has advantages and disadvantages. The choice of which one to use depends on the experimental conditions.

The radioactive nucleoside tritiated thymidine was first used to label DNA synthesis until methods that eliminated the use of radioactivity were developed. Thymidine analogs such as 5-bromo-2′-deoxyuridine are commonly used in place of radioactive thymidine. Using a thymidine analog to label DNA synthesis is reliable and accurate. It requires the injection of the analog into an animal or incubation in cells or tissue in culture where it is integrated into replicating DNA. Thymidine integration is largely permanent, present for the life of the cell, which allows one to see the fate of the new cell at later time points. Radioactive thymidine is observed using autoradiography that measures radioactive decay. Nonradioactive thymidine analogs can be visualized using reporters such as fluorescent antibodies. One of the drawbacks is that incorporation of the thymidine analog can result in detrimental effects on cell-cycle progression and growth (Anda et al., 2014).

In vitro tetrazolium salts are used to measure cellular metabolism in cultured cells by serving as a substrate that is converted by reduction into a compound with a unique absorbance wavelength measured with a spectrophotometer (Riss et al., 2004). A comparison of cells under different conditions can reveal actively dividing cells versus nondividing, viable cells. Several different tetrazolium salts are available and 3-(4,5-dimethylthiazol-2-yl)-2,5-diphenyltetrazolium bromide (MTT) readily penetrates eukaryotic cells. Some salts require a solvent such as dimethyl sulfoxide (DMSO). Additionally, the choice of salt may depend on solubility in normal culture medium and toxicity.

There are three proliferation proteins known to indicate cell division: proliferating cell nuclear antigen (PCNA), Ki67 and minichromosome maintenance complex component 2 (MCM2). All three are present during G1, S, G2, and M phases of the cell cycle, but absent during G0. Antibodies that target each protein can be used to visualize which cells are proliferating in fixed tissue sections or cultured cells. Studies have shown the efficacy of one over another for detecting certain forms of cancer suggesting that careful evaluation of each marker should be done before determining which is best for a particular study (Oka et al., 2011; Szelachowska et al., 2006).

Mesenchymal Stem Cells: Somatic Stem Cell Potency

The mesenchymal stem cells (MSCs) that arise from the mesenchymal germ layer provide an ideal paradigm to investigate some of the important outstanding questions. MSCs have been found in several different locations of the human body including cancellous (spongy) bone, adipose tissue, the umbilical cord blood and matrix (intracellular tissue of the umbilical cord), amniotic fluid, placenta, dental pulp, and tendons (New et al., 2015; Schipani and Kronenberg, 2008). The presence of MSCs in these disparate tissues is cause for comparison such as: (1) whether they have a similar identity (determined by molecular marker expression, (2) whether they have equivalent differentiation potential, (3) do they use the same signaling mechanisms to determine whether symmetric or asymmetric division will occur, and (4) whether somatic stem cells isolated from different tissues use the same signaling mechanisms to undergo division? To address the first question, initial research comparing MSCs from different locations found that they express a similar panel of *CD markers* typical of both MSCs and HSCs (Box 2.2). Surprisingly, umbilical cord-derived and pediatric adipose-derived MSCs express CD markers associated with a neural phenotype (Christodoulou et al., 2013; New et al., 2015). Further comparison found that the in vitro proliferative potential of either of these MSC sources could be manipulated by changing the culture conditions. Based on the number of somatic stem cell populations in the human body, it is clear from these studies that despite many shared characteristics, we cannot assume that all MSCs or all somatic stem cells are identical.

Intestinal Stem Cells: Regulation of Proliferation

Generally, somatic stem cells respond to signals indicating the need to replace cells or repair the tissue in which they reside. In order to maintain homeostasis some somatic stem cells must proliferate rapidly in tissues with high turnover (gut, skin), while other must maintain quiescence for longer periods of time (hair follicle stem cells). For example, continued growth of the rodent incisor is supported by slowly dividing epithelial stem cells that counteract abrasion (Harada et al., 2002). In contrast, intestinal stem cells (ISCs) are a rapidly dividing population that can replace every cell in the female *Drosophila* midgut in 2 weeks (4 weeks in males), a rate that increases upon injury or infection (Jiang et al., 2009; Micchelli and Perrimon, 2006; Ohlstein and Spradling, 2006). It is not known whether the signals for proliferation are the same in rapidly versus slowly dividing stem cell populations or under injury- or stress-induced conditions (Biteau et al., 2011). However, results from *Drosophila* studies have worked out a signaling mechanism that underlies asymmetric stem cell division, an important regulator of proliferation.

BOX 2.2

CLUSTERS OF DIFFERENTIATION

The clusters of differentiation (CD) database originated as a way to classify antigens expressed by leukocytes (white blood cells derived from HSCs) and the monoclonal antibodies that bind those antigens (Fig. 2.4). The antigens may belong to any class of protein or molecule and may have various functions such as receptor, ligand or adhesion molecules. Once an antigen is well characterized using a known list of monoclonal antibodies to which it binds, it is assigned a CD number. Tissue or cells can be characterized for antigen expression using the monoclonal antibodies, which are then visualized with a second antibody that is tagged with a fluorescent marker.

Thanks to the work of earlier research, today's investigator can look up the CD numbers to determine what type of protein or molecule it describes. For example, CD133, a marker of neural precursors, is a receptor for the Wnt-2a ligand called Frizzled-9 (Fz-9). Wnt proteins are a large ligand family; Wnt signaling contributes to many aspects of development including patterning of the embryo, organizing the nervous system and the differentiation and maturation of neurons. Thus CD number characterization can determine shared signaling pathways (including neurotransmitters), transcription factor expression, and so on between different stem cell populations, providing clues to how they function.

The CD nomenclature has been used to classify somatic stem cells types, and determine their similarity to ESCs and ESC-like stem cells. Umbilical cord-derived MSCs and pediatric adipose-derived MSCs share the same CD markers suggesting that they express the same proteins and therefore may share some of the same physiology (Christodoulou et al., 2013). Combining data from Christodoulou et al. (2013) and New et al. (2015) shows that umbilical cord-derived and pediatric adipose-derived MSCs both express CD13, 29, 44, 73, 90, 105, and 166. HSC-derived MSCs express CD14, 34, and 45, while neural stem cells express CD34, 133, and 349 (Christodoulou et al., 2013; New et al., 2015). A simple interpretation would be that mesenchymal HSCs and neural stem cells have more in common compared to HSCs and the umbilical cord-derived and pediatric adipose-derived MSCs. However, we cannot make that assumption without more evidence. CD markers can provide the starting point for a hypothesis, but further research into the differences between stem cell types is necessary.

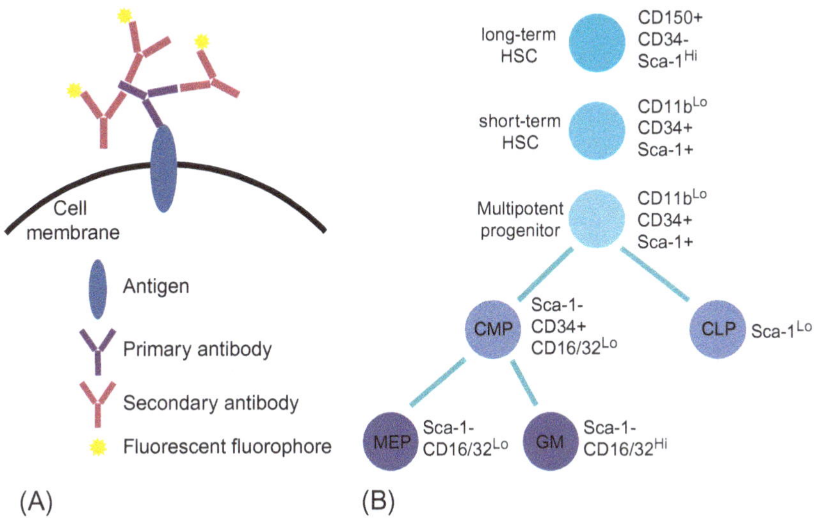

FIG. 2.4 Clusters of differentiation. Antibodies are proteins that recognize and bind to specific antigens expressed by cells. Primary monoclonal antibodies made from a single source bind to and label only one antigen. Secondary antibodies conjugated with a fluorescent fluorophore recognize and bind to general features on the primary antibodies (A). When a secondary antibody binds a primary antibody on a cell or in a tissue, the fluorescence indicates the presence of the antigen. This technology was used to study and classify the different types of white blood cells. During the 1980s scientists produced multiple primary antibodies that recognized different antigens on several white blood cell types. A large workshop was held to determine which antibodies labeled each of the white blood cell types (Zola et al., 2005). Groups of antibodies that labeled the same cell type were given a cluster number. Now the CD nomenclature can refer to proteins expressed by white blood cells or the antibodies that bind to them. (B)shows a simplified version of the markers that label hematopoietic stem and progenitor cells (Pazhanisamy, 2013). *HSC*, hematopoietic stem cell; *CLP*, common lymphoid progenitor; *CMP*, common myeloid progenitor; *MEP*, megakaryocytes and erythroid cells; *GM*, granulocytes and macrophages. *(B) Modified with permission from Mater Methods, 2013.*

The *Drosophila* gut consists of four parts: the foregut, midgut, Malpighian tubules and the hindgut (Fig. 2.5A; Buchon et al., 2013). The midgut is the equivalent of the mammalian small intestine and is of endodermal origin. It consists of an epithelial layer of columnar cells that are insulated from gut bacteria by the peritrophic membrane and separated from visceral muscles by a basement membrane (Fig. 2.5B). ISCs within the midgut epithelia give rise to the two mature cell types in the gut: nutrient absorbing enterocytes (ECs) and digestive hormone secretory enteroendocrine cells (EEs). When ISCs divide symmetrically the result is either two new ISCs or two differentiated, committed diploid progenitors called enteroblasts (EBs) (de Navascués et al., 2012). When ISCs divide asymmetrically the result is one new ISC and one EB committed to differentiation, becoming either an EC or an EE.

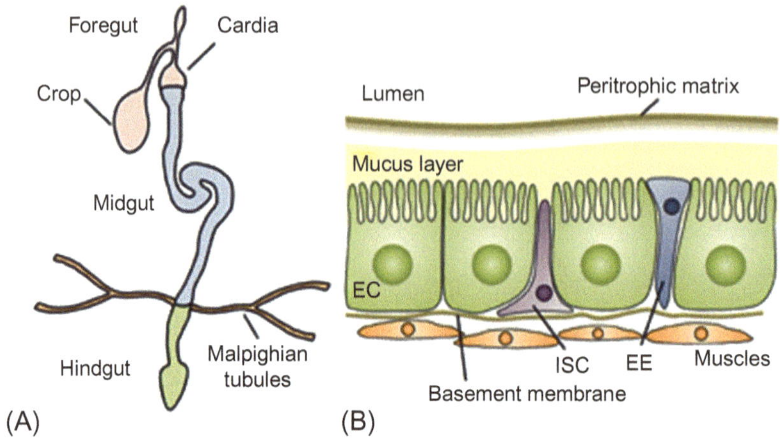

FIG. 2.5 The adult *Drosophila* midgut contains intestinal stem cells (ISCs). (A) A diagram of the adult *Drosophila* gut. Malpighian tubes are the equivalent of the mammalian kidneys. (B) A cross-section of the adult midgut shows multipotent ISCs (purple) at the base of the enterocytes (EC, green) adjacent to the basement membrane and the visceral muscle (orange). ISCs give rise to two types of differentiated daughters, enteroendocrine (EE) cells (blue) and ECs. *Copyright © 2013 Kuraishi, T., Hori, A., Kurata, S., 2013. Host-microbe interactions in the gut of* Drosophila melanogaster. *Front. Physiol. 4. This is an open-access article distributed under the terms of the Creative Commons Attribution License (CC BY).*

Whether it is the fate of an EB cell to differentiate into an EE or an EC cell depends on the Notch signaling pathway (Fig. 2.6). Notch is the name of a transmembrane receptor with an intracellular and extracellular domain; its ligand is called Delta. Delta is a transmembrane protein expressed on the cell surface and requires contact with a cell expressing the Notch receptor to initiate intracellular signaling. After an asymmetric division only the new ISC expresses the Delta ligand and activates Notch signaling in the EB cell (Vooijs et al., 2011). In 90% of the ISCs expression of Delta is high and strongly activates Notch signaling in the EB cell, causing them to differentiate into EC cells (Ohlstein and Spradling, 2007). In 10% of the ISCs, low Delta expression weakly activates Notch signaling in the EB cells, which causes them to differentiate into EE cells.

After the initial cell division the daughter cell in greater contact with the basement membrane has higher Delta expression and self-renews as an ISC through repression of a group of *transcription factor* genes called the *Enhancer of split* complex (*E*(*spl*)-C) (Ohlstein and Spradling, 2006). The more apically located daughter cell has lower Delta expression such that Delta in the adjacent ISC activates Notch signaling resulting in an EB fate. Delta binding to Notch in the EB daughter cell triggers the proteolytic cleavage of Notch. The cleaved intracellular domain of Notch enters the cell nucleus where it makes a complex with a transcription factor called Suppressor of Hairless (Su(H)). Su(H) affects two basic helix-loop-helix

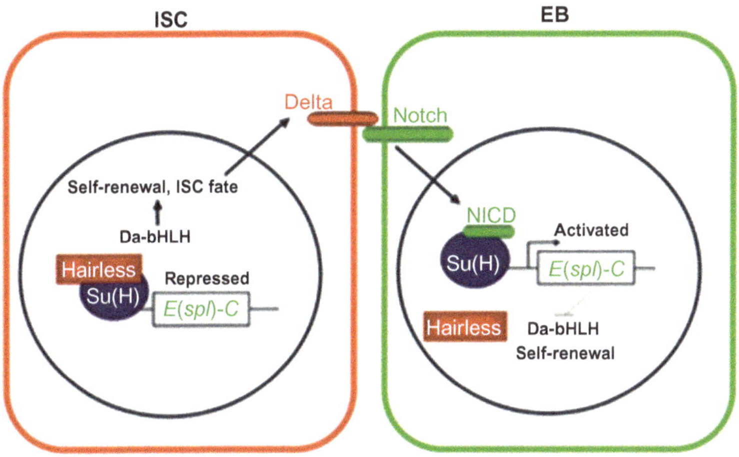

FIG. 2.6 In the ISC expression of the basic helix-loop-helix transcription factor called Daughterless (Da-bHLH) maintains a stem cell fate while Hairless suppresses genes that would promote differentiation. When Delta ligands binds to Notch receptors on the new daughter cell, an EB fate is promoted. The intracellular domain of Notch (NICD) forms a complex with Suppressor of Hairless (Su(H) that activates expression of a group of transcription factor genes called the Enhancer of split complex (E(spl)-C), which represses Da-bHLH. This sequence of events promotes differentiation to an EE or EC cell fate. *From Bardin, A.J., Perdigoto, C.N., Southall, T.D., Brand, A.H., Schweisguth, F., 2010. Transcriptional control of stem cell maintenance in the Drosophila intestine. Development 137, 705–714.*

transcription factors: it activates *E(spl)*-C, which represses Daughterless (Da) (Bardin et al., 2010). This sequence of events prevents ISC identity and promotes differentiation to an EE or more common EC cell fate.

Biliary Tree Stem Cells: Discovery and Application

It is still not known whether we have discovered all the somatic stem cell types within the human body. For example, as recently as 2009 a new source of stem cells was confirmed in the biliary tree of the liver that has the potential to treat Type I diabetes (De Alwis et al., 2009) (Box 2.3). The biliary tree is a structure composed of interconnected bile ducts that branch extensively throughout the liver (intrahepatic) and connects to the gall bladder and pancreas (extrahepatic). The intrahepatic bile ducts within the liver are called canals of Hering. Several stem cell niches were found in the canals of Hering where the recently discovered biliary tree stem cells reside (BTSCs; Fig. 2.7).

The BTSCs are multipotent stem/progenitor cells that grow rapidly in culture, dividing every ~36 h and slowing to one division every 2–3 days while remaining undifferentiated for months. BTSCs express stem/progenitor surface markers, giving rise to at least two confirmed

BOX 2.3

SPOTLIGHT ON DIABETES

Diabetes mellitus is the most common metabolic disorder, currently affecting more than 29 million Americans and one in four persons over 65 years old, according to the American Diabetes Association. Type 1 diabetes is an autoimmune disorder in which the body's immune system attacks the pancreatic beta (β) cells that produce insulin. Type 2 diabetes is acquired, but can usually be prevented by a healthy lifestyle. Several studies show some promise in using stem cells to treat both diabetic types. The current gold standard treatment for Type 1 diabetes is a whole pancreatic transplant. Since the first pancreatic transplant surgery in 1966 more than 30,000 patients have received a pancreas. Pancreatic islet cell transplantations are another option, but immune rejection and a shortage of donors has limited the success of this therapy. Thus stem cells have received much attention as a potential source of new β cells and new hope for diabetes sufferers.

To obtain good metabolic control by production of insulin it takes approximately 2×10^6 β cells per kg of patient body weight (Keymeulen et al., 2006). One research group developed a way to mass produce β cells using a line of human ESCs and two human-induced pluripotent stem cell lines (Pagliuca et al., 2014). Using an in vitro culture system, billions of β cells were produced without genetic modification or the addition of mesenchymal or endothelial cells that could contaminate (either cellularly or genetically) the final product. The protocol took 4–5 weeks and required the sequential addition of factors such as small molecule agonists and antagonists and assorted growth factors that affect signaling in numerous pathways that determine the differentiation of stem cells into adult pancreatic β cells (Pagliuca et al., 2014).

The cultured β cells secreted insulin in response to glucose stimulation both in culture and when transplanted into immunocompromised mice. Next, the β cells were transplanted into a strain of immunocompromised mice that spontaneously develop hyperglycemia. The mice that received the transplanted β cells were restored to normoglycemia and had increased survival times compared to control mice (Pagliuca et al., 2014). This strategy shows great promise for Type 1 diabetes treatment, however the cultured β cells have some genetic differences from adult human β cells, and long-term assessment is required before a clinical trial can be carried out.

A major hurdle to stem cell clinical therapy is immune system rejection. Current research is exploring the use of an encapsulation device that

BOX 2.3

SPOTLIGHT ON DIABETES—CONT'D

contains isolated allogeneic (immunologically incompatible) human islets formulated in a serum-free transplant media that can be implanted subcutaneously (Domínguez-Bendala et al., 2016). The device allows oxygen and nutrients to enter the device, but prevents direct contact with immune cells (Boettler et al., 2016; Kirk et al., 2014). A study using primates suggests that encapsulation successfully hides the islets from the immune system, while allowing insulin produced by the islet cells to exit the device (Vériter et al., 2014). Due to a number of factors, this technology is not yet ready as a treatment for diabetes (Krishnan et al., 2014). One of the improvements to this device is to include MSCs within the capsule, which increased the islet cells' survival (Davis et al., 2012). Innovative technology and an increased understanding of stem cell physiology have the potential to make a Type 1 diabetes diagnosis much easier to bear.

transit-amplifying cells, a type of intermediate progenitor with reduced potency that then give rise to committed progenitors (Cardinale et al., 2011, 2012; Wang et al., 2010). By testing distinct culture conditions it was found that a hormonally defined medium encouraged the growth of pancreatic islet-like cells (neoislets). Pancreatic islets are clusters of 3000–4000 cells including β cells that sense glucose in the blood and respond by releasing insulin. The neoislets expressed markers correlating with adult pancreatic islets (Cardinale et al., 2011). To test effectiveness in vivo, BTSCs were transplanted into immunosuppressed mice that had been rendered diabetic via chemical treatment (streptozocin-induced hyperglycemia). The transplanted BTSCs allowed the mice to recover from hyperglycemia over a two-month period. Insulin levels in the BTSC-treated mice were regulatable by glucose, a characteristic of adult pancreatic islets. These data suggest that BTSCs may be a potential treatment for Type 1 diabetes. Findings such as these continue to drive stem cell research.

A comparison of embryonic and somatic stem cells:

ESCs and somatic stem cells each have advantages over the other. ESCs are easily grown in culture where they proliferate robustly. Somatic stem cells are difficult to obtain in large numbers from the tissue or organ in which they reside and typically do not proliferate rapidly in culture. However, somatic stem cells isolated from a patient that are eventually put back in the patient will be recognized as "self." Transplanted ESCs are rejected unless the host is either immunodeficient or taking immune

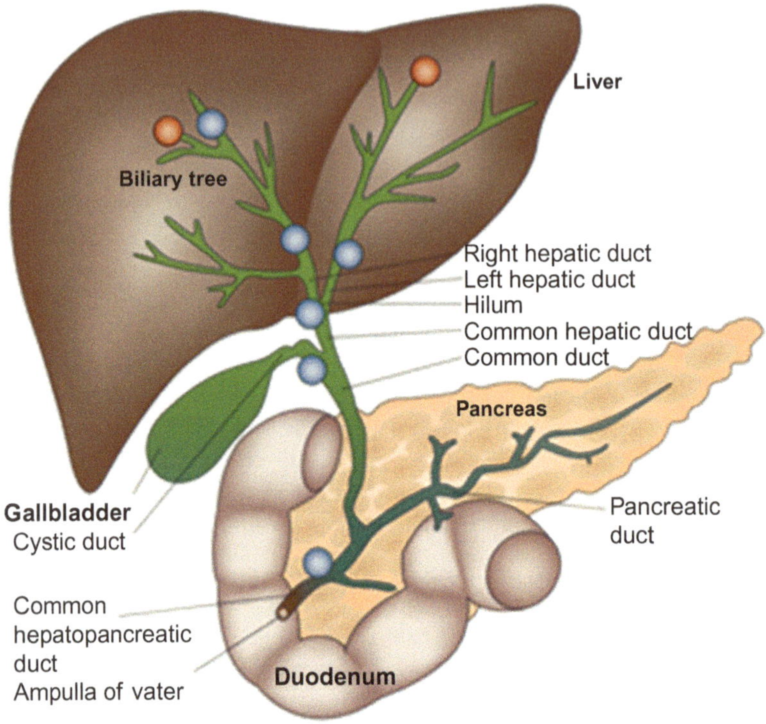

FIG. 2.7 The biliary tree consists of bile ducts, including the gall bladder, that branch extensively into the liver (intrahepatic) and pancreas (extrahepatic). In the biliary tree branches several stem cell niches are present in both intra- and extrahepatic locations (blue circles). Stem cell niches have also been found in the intrahepatic bile ducts called canals of Hering (red circles). *From Cardinale, V., Wang, Y., Carpino, G., Mendel, G., Alpini, G., Gaudio, E., Reid, L.M., Alvaro, D., 2012. The biliary tree—a reservoir of multipotent stem cells. Nat. Rev. Gastroenterol. Hepatol. 9, 231–240.*

suppression drugs. ESCs also have ethical issues associated with using them for research as discussed in Chapter 1.

ESCs were thought to be the most potent cell type, but scientists are finding ways to *dedifferentiate* somatic cells. By controlling gene expression, fate-determining genes are turned off, while genes necessary for self-renewal and pluripotency are turned on to induce a somatic cell into a pluripotent cell. Thus somatic cells are now being studied for their potential to give rise to cells outside the germ layer from which they originated through dedifferentiation followed by controlled differentiation. In time, patients may provide their own cells that could be manipulated into treatments (more on this in Chapter 4).

SUMMARY

Our understanding of human and cellular development has led to new discoveries of embryonic-like and somatic stem cell sources. Common properties have been found in embryonic and somatic stem cells, such as self-renewal and multipotency. Current research is examining the similarities and differences between these cell types. In vitro, embryonic and somatic stem cells behave differently. In vivo, somatic stem cells in different locations of the body may have very different proliferation rates depending on the tissue or organ in which they are expressed. However, studies continue to determine how proliferation is regulated at each location and whether common signaling pathways are a shared feature. It is essential to understand how stem cells proliferate—and how to control it—as there is a concern that transplanted stem cells may become tumor forming. Answering these basic research questions will enable further progress on stem cells for regenerative therapy.

DEFINITIONS

Asymmetric cell division A stem cell divides to produce one identical copy of itself (self-renewal) and one cell that is different from the mother cell, typically more committed to becoming a specific cell type than the original.

CD markers Clusters of differentiation markers. A defined set of antibodies known to label a specific antigen.

Dedifferentiate To reverse the state of a cell from a determined developmental fate with specialized characteristics to an undifferentiated multipotent or pluripotent state.

Embryonic stem cell Pluripotent stem cells present during the blastocyst stage of embryo development. They are self-renewing cells and have the potential to become any cell in the body.

Generative asymmetric cell division A stem cell produces one identical copy of itself (self-renewal) and one cell that is specialized. In the brain this occurs over a long period to generate neural cells (neurons and glia).

Generative symmetric cell division At the end of development stem cells may switch from asymmetric to symmetric cell division and generate two specialized cells, thus reducing the stem cell population.

Multipotent The potential to become any cell type *within* a germ layer, organ or tissue.

Niche A specialized microenvironment in which stem cells reside. The niche provides a location for stem cells and/or factors that influence the fate of a dividing stem cell. For example, neighboring cells

	may secrete factors that increase or decrease stem cell proliferation.
Proliferative symmetric cell division	A stem cell divides to produce two identical copies of itself.
Somatic stem cells	A multi or unipotent stem cell found in adult tissues or organs throughout the body that provides lifelong replacement, replenishment or repair of specific cells.
Symmetric cell division (self-renewal)	When a stem cell divides to produce two new copies of itself. This is a defining feature of stem cells.
Totipotent	A cell with the ability to produce every cell in the body *and* the placenta. Totipotency is the initial state of a fertilized embryo.
Transit-amplifying cells	Cells that arise directly from stem cells that proliferate for a finite period of time and then differentiate.
Transcription factors	Proteins that bind to or interact with DNA to produce RNA. The part of the transcription factor that interacts directly with the DNA is called the homeodomain, typically about 60 amino acids long. Transcription factors have several conserved structural motifs that facilitate interaction with DNA such as the basic helix-loop-helix motif.
Unipotent	The potential to produce a single cell type in addition to its own self-renewal.

References

Abdulrazzak, H., De Coppi, P., Guillot, P.V., 2013. Therapeutic potential of amniotic fluid stem cells. Curr. Stem Cell Res. Ther. 8, 117–124.

Abramson, S., Miller, R.G., Phillips, R.A., 1977. The identification in adult bone marrow of pluripotent and restricted stem cells of the myeloid and lymphoid systems. J. Exp. Med. 145, 1567–1579.

Altman, J., Das, G.D., 1965. Autoradiographic and histological evidence of postnatal hippocampal neurogenesis in rats. J. Comp. Neurol. 124, 319–335.

Anda, S., Boye, E., Grallert, B., 2014. Cell-cycle analyses using thymidine analogues in fission yeast. PLoS ONE 9, e88629.

Bardin, A.J., Perdigoto, C.N., Southall, T.D., Brand, A.H., Schweisguth, F., 2010. Transcriptional control of stem cell maintenance in the Drosophila intestine. Development 137, 705–714.

Biteau, B., Hochmuth, C.E., Jasper, H., 2011. Maintaining tissue homeostasis: dynamic control of somatic stem cell activity. Cell Stem Cell 9, 402–411.

Boettler, T., Schneider, D., Cheng, Y., Kadoya, K., Brandon, E.P., Martinson, L., Von Herrath, M., 2016. Pancreatic tissue transplanted in TheraCyte ™ encapsulation devices is protected and prevents hyperglycemia in a mouse model of immune-mediated diabetes. Cell Transplant. 25, 609–614.

Bonaventura, G., Chamayou, S., Liprino, A., Guglielmino, A., Fichera, M., Caruso, M., Barcellona, M.L., 2015. Different tissue-derived stem cells: a comparison of neural differentiation capability. PLoS One 10, e0140790.

Buchon, N., Osman, D., David, F.P.A., Yu Fang, H., Boquete, J.-P., Deplancke, B., Lemaitre, B., 2013. Morphological and molecular characterization of adult midgut compartmentalization in Drosophila. Cell Rep. 3, 1725–1738.

Cardinale, V., Wang, Y., Carpino, G., Cui, C.-B., Gatto, M., Rossi, M., Berloco, P.B., Cantafora, A., Wauthier, E., Furth, M.E., et al., 2011. Multipotent stem/progenitor cells in human biliary tree give rise to hepatocytes, cholangiocytes, and pancreatic islets. Hepatol. Baltim. Md. 54, 2159–2172.

Cardinale, V., Wang, Y., Carpino, G., Mendel, G., Alpini, G., Gaudio, E., Reid, L.M., Alvaro, D., 2012. The biliary tree—a reservoir of multipotent stem cells. Nat. Rev. Gastroenterol. Hepatol. 9, 231–240.

Carraro, G., Perin, L., Sedrakyan, S., Giuliani, S., Tiozzo, C., Lee, J., Turcatel, G., De Langhe, S.P., Driscoll, B., Bellusci, S., et al., 2008. Human amniotic fluid stem cells can integrate and differentiate into epithelial lung lineages. Stem Cells Dayt. Ohio 26, 2902–2911.

Christodoulou, I., Kolisis, F.N., Papaevangeliou, D., Zoumpourlis, V., 2013. Comparative evaluation of human mesenchymal stem cells of fetal (Wharton's jelly) and adult (adipose tissue) origin during prolonged in vitro expansion: considerations for cytotherapy. Stem Cells Int. 2013, 246134.

Davis, N.E., Beenken-Rothkopf, L.N., Mirsoian, A., Kojic, N., Kaplan, D.L., Barron, A.E., Fontaine, M.J., 2012. Enhanced function of pancreatic islets co-encapsulated with ECM proteins and mesenchymal stromal cells in a silk hydrogel. Biomaterials 33, 6691–6697.

De Alwis, N., Hudson, G., Burt, A.D., Day, C.P., Chinnery, P.F., 2009. Human liver stem cells originate from the canals of Hering. Hepatol. Baltim. Md. 50, 992–993.

De Coppi, P., Callegari, A., Chiavegato, A., Gasparotto, L., Piccoli, M., Taiani, J., Pozzobon, M., Boldrin, L., Okabe, M., Cozzi, E., et al., 2007a. Amniotic fluid and bone marrow derived mesenchymal stem cells can be converted to smooth muscle cells in the cryo-injured rat bladder and prevent compensatory hypertrophy of surviving smooth muscle cells. J. Urol. 177, 369–376.

De Coppi, P., Bartsch, G., Siddiqui, M.M., Xu, T., Santos, C.C., Perin, L., Mostoslavsky, G., Serre, A.C., Snyder, E.Y., Yoo, J.J., et al., 2007b. Isolation of amniotic stem cell lines with potential for therapy. Nat. Biotechnol. 25, 100–106.

de Navascués, J., Perdigoto, C.N., Bian, Y., Schneider, M.H., Bardin, A.J., Martínez-Arias, A., Simons, B.D., 2012. *Drosophila* midgut homeostasis involves neutral competition between symmetrically dividing intestinal stem cells: Drosophila midgut homeostasis involves ISC neutral competition. EMBO J. 31, 2473–2485.

De Paepe, C., Krivega, M., Cauffman, G., Geens, M., Van de Velde, H., 2014. Totipotency and lineage segregation in the human embryo. MHR Basic Sci. Reprod. Med. 20, 599–618.

Denner, L., Bodenburg, Y., Zhao, J.G., Howe, M., Cappo, J., Tilton, R.G., Copland, J.A., Forraz, N., McGuckin, C., Urban, R., 2007. Directed engineering of umbilical cord blood stem cells to produce C-peptide and insulin. Cell Prolif. 40, 367–380.

Domínguez-Bendala, J., Lanzoni, G., Klein, D., Álvarez-Cubela, S., Pastori, R.L., 2016. The human endocrine pancreas: new insights on replacement and regeneration. Trends Endocrinol. Metab. 27, 153–162.

Fauza, D., 2004. Amniotic fluid and placental stem cells. Best Pract. Res. Clin. Obstet. Gynaecol. 18, 877–891.

Fico, A., de Chevigny, A., Melon, C., Bohic, M., Kerkerian-Le Goff, L., Maina, F., Dono, R., Cremer, H., 2014. Reducing glypican-4 in ES cells improves recovery in a rat model of Parkinson's disease by increasing the production of dopaminergic neurons and decreasing teratoma formation. J. Neurosci. 34, 8318–8323.

Forraz, N., Pettengell, R., Deglesne, P.-A., McGuckin, C.P., 2002. AC133+ umbilical cord blood progenitors demonstrate rapid self-renewal and low apoptosis. Br. J. Haematol. 119, 516–524.

Gardner, D.K., Lane, M., 1998. Culture of viable human blastocysts in defined sequential serum-free media. Hum. Reprod. 13, 148–159.

Harada, H., Mitsuyasu, T., Toyono, T., Toyoshima, K., 2002. Epithelial stem cells in teeth. Odontology 90, 1–6.

Harris, D.T., Rogers, I., 2007. Umbilical cord blood: a unique source of pluripotent stem cells for regenerative medicine. Curr. Stem Cell Res. Ther. 2, 301–309.

He, J.-Q., Ma, Y., Lee, Y., Thomson, J.A., Kamp, T.J., 2003. Human embryonic stem cells develop into multiple types of cardiac myocytes: action potential characterization. Circ. Res. 93, 32–39.

Huttner, W.B., Kosodo, Y., 2005. Symmetric versus asymmetric cell division during neurogenesis in the developing vertebrate central nervous system. Curr. Opin. Cell Biol. 17, 648–657.

In 't Anker, P.S., Scherjon, S.A., Kleijburg-van der Keur, C., Noort, W.A., Claas, F.H.J., Willemze, R., Fibbe, W.E., Kanhai, H.H.H., 2003. Amniotic fluid as a novel source of mesenchymal stem cells for therapeutic transplantation. Blood 102, 1548–1549.

Jiang, H., Patel, P.H., Kohlmaier, A., Grenley, M.O., McEwen, D.G., Edgar, B.A., 2009. Cytokine/Jak/Stat signaling mediates regeneration and homeostasis in the Drosophila midgut. Cell 137, 1343–1355.

Kaltschmidt, J.A., Davidson, C.M., Brown, N.H., Brand, A.H., 2000. Rotation and asymmetry of the mitotic spindle direct asymmetric cell division in the developing central nervous system. Nat. Cell Biol. 2, 7–12.

Kaufman, D.S., Hanson, E.T., Lewis, R.L., Auerbach, R., Thomson, J.A., 2001. Hematopoietic colony-forming cells derived from human embryonic stem cells. Proc. Natl. Acad. Sci. U. S. A. 98, 10716–10721.

Keymeulen, B., Gillard, P., Mathieu, C., Movahedi, B., Maleux, G., Delvaux, G., Ysebaert, D., Roep, B., Vandemeulebroucke, E., Marichal, M., et al., 2006. Correlation between beta cell mass and glycemic control in type 1 diabetic recipients of islet cell graft. Proc. Natl. Acad. Sci. 103, 17444–17449.

Kimble, J., 2014. *C. elegans* germline stem cells and their niche. StemBook.

Kirk, K., Hao, E., Lahmy, R., Itkin-Ansari, P., 2014. Human embryonic stem cell derived islet progenitors mature inside an encapsulation device without evidence of increased biomass or cell escape. Stem Cell Res. 12, 807–814.

Klimanskaya, I., Chung, Y., Meisner, L., Johnson, J., West, M.D., Lanza, R., 2005. Human embryonic stem cells derived without feeder cells. Lancet 365, 1636–1641.

Knoblich, J.A., 2008. Mechanisms of asymmetric stem cell division. Cell 132, 583–597.

Kolambkar, Y.M., Peister, A., Soker, S., Atala, A., Guldberg, R.E., 2007. Chondrogenic differentiation of amniotic fluid-derived stem cells. J. Mol. Histol. 38, 405–413.

Krishnan, R., Alexander, M., Robles, L., Foster, C.E., Lakey, J.R.T., 2014. Islet and stem cell encapsulation for clinical transplantation. Rev. Diabet. Stud. 11, 84–101.

Lin, H., Schagat, T., 1997. Neuroblasts: a model for the asymmetric division of stem cells. Trends Genet. TIG 13, 33–39.

McGuckin, C.P., Forraz, N., Allouard, Q., Pettengell, R., 2004. Umbilical cord blood stem cells can expand hematopoietic and neuroglial progenitors in vitro. Exp. Cell Res. 295, 350–359.

McGuckin, C., Jurga, M., Ali, H., Strbad, M., Forraz, N., 2008. Culture of embryonic-like stem cells from human umbilical cord blood and onward differentiation to neural cells in vitro. Nat. Protoc. 3, 1046–1055.

Micchelli, C.A., Perrimon, N., 2006. Evidence that stem cells reside in the adult Drosophila midgut epithelium. Nature 439, 475–479.

Moon, Y.J., Lee, M.W., Yoon, H.H., Yang, M.S., Jang, I.K., Lee, J.E., Kim, H.E., Eom, Y., Park, J.S., Kim, H.C., et al., 2008. Hepatic differentiation of cord blood-derived multipotent progenitor cells (MPCs) in vitro. Cell Biol. Int. 32, 1293–1301.

Morrison, S.J., Kimble, J., 2006. Asymmetric and symmetric stem-cell divisions in development and cancer. Nature 441, 1068–1074.

New, S.E.P., Alvarez-Gonzalez, C., Vagaska, B., Gomez, S.G., Bulstrode, N.W., Madrigal, A., Ferretti, P., 2015. A matter of identity—phenotype and differentiation potential of human somatic stem cells. Stem Cell Res. 15, 1–13.

Ohlstein, B., Spradling, A., 2006. The adult Drosophila posterior midgut is maintained by pluripotent stem cells. Nature 439, 470–474.

Ohlstein, B., Spradling, A., 2007. Multipotent Drosophila intestinal stem cells specify daughter cell fates by differential notch signaling. Science 315, 988–992.

Oka, S., Uramoto, H., Shimokawa, H., Iwanami, T., Tanaka, F., 2011. The expression of Ki-67, but not proliferating cell nuclear antigen, predicts poor disease free survival in patients with adenocarcinoma of the lung. Anticancer Res. 31, 4277–4282.

Pagliuca, F.W., Millman, J.R., Gürtler, M., Segel, M., Van Dervort, A., Ryu, J.H., Peterson, Q.P., Greiner, D., Melton, D.A., 2014. Generation of functional human pancreatic β cells in vitro. Cell 159, 428–439.

Pazhanisamy, S., 2013. Adult stem cell and embryonic stem cell markers. Mater. Methods 3.

Perin, L., Giuliani, S., Jin, D., Sedrakyan, S., Carraro, G., Habibian, R., Warburton, D., Atala, A., De Filippo, R.E., 2007. Renal differentiation of amniotic fluid stem cells. Cell Prolif. 40, 936–948.

Riss, T., Moravec, R., Niles, A., Duellman, S., Benink, H., Worzella, M., Minor, L., 2004. Cell viability assays. In: Assay Guidance Manual. Eli Lilly & Company and the National Center for Advancing Translational Science, Bethesda, MD.

Rossant, J., Tam, P.P.L., 2009. Blastocyst lineage formation, early embryonic asymmetries and axis patterning in the mouse. Development 136, 701–713.

Saiz, N., Plusa, B., 2013. Early cell fate decisions in the mouse embryo. Reproduction 145, R65–R80.

Schipani, E., Kronenberg, H.M., 2008. Adult mesenchymal stem cells. In: StemBook. Harvard Stem Cell Institute, Cambridge, MA.

Stevens, L.C., 1970. The development of transplantable teratocarcinomas from intratesticular grafts of pre- and postimplantation mouse embryos. Dev. Biol. 21, 364–382.

Sun, H., Feng, K., Hu, J., Soker, S., Atala, A., Ma, P.X., 2010. Osteogenic differentiation of human amniotic fluid-derived stem cells induced by bone morphogenetic protein-7 and enhanced by nanofibrous scaffolds. Biomaterials 31, 1133–1139.

Szelachowska, J., Dziegiel, P., Jelen-Krzeszewska, J., Jelen, M., Matkowski, R., Pomiecko, A., Spytkowska, B., Jagas, M., Gisterek, I., Kornafel, J., 2006. Mcm-2 protein expression predicts prognosis better than Ki-67 antigen in oral cavity squamocellular carcinoma. Anticancer Res. 26, 2473–2478.

Tamplin, O.J., Durand, E.M., Carr, L.A., Childs, S.J., Hagedorn, E.J., Li, P., Yzaguirre, A.D., Speck, N.A., Zon, L.I., 2015. Hematopoietic stem cell arrival triggers dynamic remodeling of the perivascular niche. Cell 160, 241–252.

Thomson, J.A., Kalishman, J., Golos, T.G., Durning, M., Harris, C.P., Becker, R.A., Hearn, J.P., 1995. Isolation of a primate embryonic stem cell line. Proc. Natl. Acad. Sci. U. S. A. 92, 7844–7848.

Thomson, J.A., Itskovitz-Eldor, J., Shapiro, S.S., Waknitz, M.A., Swiergiel, J.J., Marshall, V.S., Jones, J.M., 1998. Embryonic stem cell lines derived from human blastocysts. Science 282, 1145–1147.

Vériter, S., Gianello, P., Igarashi, Y., Beaurin, G., Ghyselinck, A., Aouassar, N., Jordan, B., Gallez, B., Dufrane, D., 2014. Improvement of subcutaneous bioartificial pancreas vascularization and function by coencapsulation of pig islets and mesenchymal stem cells in primates. Cell Transplant. 23, 1349–1364.

Visvader, J.E., Clevers, H., 2016. Tissue-specific designs of stem cell hierarchies. Nat. Cell Biol. 18, 349–355.

Vooijs, M., Liu, Z., Kopan, R., 2011. Notch: architect, landscaper, and guardian of the intestine. Gastroenterology 141, 448–459.

Wang, Y., Yao, H.-L., Cui, C.-B., Wauthier, E., Barbier, C., Costello, M.J., Moss, N., Yamauchi, M., Sricholpech, M., Gerber, D., et al., 2010. Paracrine signals from mesenchymal cell populations govern the expansion and differentiation of human hepatic stem cells to adult liver fates. Hepatol. Baltim. Md. 52, 1443–1454.

Witkowska-Zimny, M., Wrobel, E., 2011. Perinatal sources of mesenchymal stem cells: Wharton's jelly, amnion and chorion. Cell. Mol. Biol. Lett. 16.

Yamashita, Y.M., 2003. Orientation of asymmetric stem cell division by the APC tumor suppressor and centrosome. Science 301, 1547–1550.
Yamashita, Y.M., Mahowald, A.P., Perlin, J.R., Fuller, M.T., 2007. Asymmetric inheritance of mother versus daughter centrosome in stem cell division. Science 315, 518–521.
Zhang, S.C., Wernig, M., Duncan, I.D., Brüstle, O., Thomson, J.A., 2001. In vitro differentiation of transplantable neural precursors from human embryonic stem cells. Nat. Biotechnol. 19, 1129–1133.
Zola, H., Swart, B., Nicholson, I., Aasted, B., Bensussan, A., Boumsell, L., Buckley, C., Clark, G., Drbal, K., Engel, P., et al., 2005. CD molecules 2005: human cell differentiation molecules. Blood 106, 3123–3126.

Further Reading

Dörner, J., Martinez Rodriguez, V., Ziegler, R., Röhrig, T., Cochran, R.S., Götz, R.M., Levin, M.D., Pihlajoki, M., Heikinheimo, M., Wilson, D.B., 2017. GLI1 + progenitor cells in the adrenal capsule of the adult mouse give rise to heterotopic gonadal-like tissue. Mol. Cell. Endocrinol. 441, 164–175.
Kuraishi, T., Hori, A., Kurata, S., 2013. Host-microbe interactions in the gut of *Drosophila melanogaster*. Front. Physiol. 4.

CHAPTER

3

The Stem Cell Niche

OUTLINE

LEARNING OBJECTIVES

1. Explain the concept of a stem cell niche.
2. Describe how a single cell can be a stem cell niche.
3. Describe the components and interactions of the Delta/Notch signaling pathway.

https://doi.org/10.1016/B978-0-12-811014-0.00003-2

4. Discuss factors inside and outside of the niche that can influence the rate of stem cell proliferation.
5. Describe the components and interactions of the JAK/STAT signaling pathway.
6. Explain the importance of quiescence as a stem cell behavior.
7. Describe the components and interactions of the Wnt signaling pathway.
8. Describe the components and interactions of the CXCL12/CXCR4 signaling pathway.
9. Explain how multiple cells expressing the CXCL12 ligand contribute to hematopoietic stem cell behavior.

INTRODUCTION

The concept of a stem cell niche was originally hypothesized by Raymond Schofield in 1978. Dr. Schofield suggested that a microenvironment exists that supports the maintenance of somatic stem cells and the formation of stem cell colonies (Schofield, 1978). Since then a diverse array of stem cell niches have been discovered in different tissue types. A stem cell niche can consist of a single cell or several cell populations that provide a location for stem cells to exist and/or regulate/influence stem cell behavior (Fig. 3.1). Niche cells may originate from outside the stem cell lineage or be derived from the stem cell lineage. A niche can be made up of cells derived from a single germ layer or of cells derived from more than one germ layer.

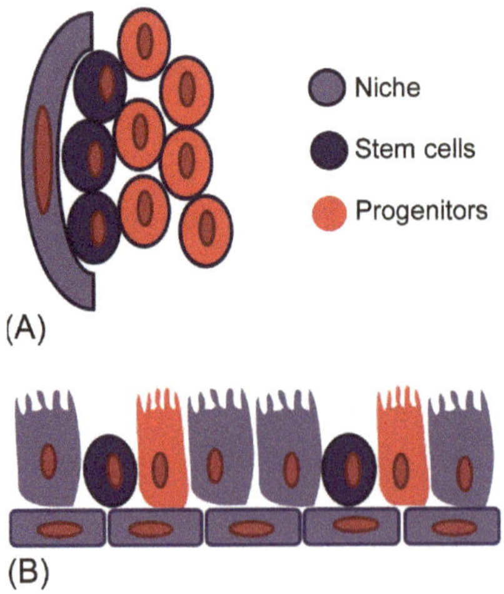

FIG. 3.1 The niche supports and influences the behavior of stem cells. The stem cell niche can consist of (A) a single cell, or (B) populations of cells surrounding the stem cell.

FUNCTIONS OF THE STEM CELL NICHE

The niche provides signals that maintain the stem cell population by preventing premature differentiation and controlling proliferation. Proliferation may be modified by homeostasis, periods of quiescence, or upregulated in response to injury or infection. Niche regulation may happen through short-range signals from neighboring cells or long-range signals from the organ in which it resides (Lay et al., 2016). Several intercellular signaling pathways have emerged that are important for regulating stem cell behaviors. While a given signaling pathway may be associated with a specific behavior, such as proliferation in more than one stem cell type, it is yet to be determined whether a given signaling pathway controls that behavior in all stem cell types. The following examples provide a glimpse of different types of somatic stem cell niches and signaling pathways that regulate behaviors, such as maintenance and proliferation.

NICHE EXAMPLES

Single-Cell Niche: The Distal Tip Cell Niche of *Caenorhabditis elegans* Germline Stem Cells

The simplest stem cell niche consists of a single cell—the distal tip cell (DTC) of the gonads in the nematode worm *C. elegans*. Of the two *C. elegans* sexes, one is male and the other is hermaphrodite. The hermaphrodite produces sperm gametes during the larval stage and oocytes as an adult. The beauty of studying *C. elegans* is that each cell of its anatomy is characterized (White et al., 1986). Furthermore, the lineage of each cell is known, beginning with the zygote all the way through larval development to maturity (Sulston et al., 1983). The diploid germline stem cells (GSCs) only give rise to *haploid* gametes but are considered totipotent because the gametes give rise to sperm and oocytes (the hermaphrodite produces both), which eventually fuse to create every cell in a new worm (Kimble, 2014).

The DTC resides in the gonad, which in a *C. elegans* hermaphrodite consists of two branch-like arms with the DTC nucleus at the end of each arm (Fig. 3.2). In the male *C. elegans*, two DTCs reside in one of the two gonads. The DTC extends long processes, giving the cell a jellyfish-like appearance that surrounds the GSCs (Kimble, 2014). A GSC maturation gradient exists based on proximity to the DTC with the GSCs closest to the DTC being totipotent and at the proximal end differentiating into gametes. The GSCs are self-renewing via symmetric mitosis near the DTC, but cells that divide further from the DTC commit to a mitotic transit-amplifying state, only proliferating for a brief period, from which they gradually differentiate to

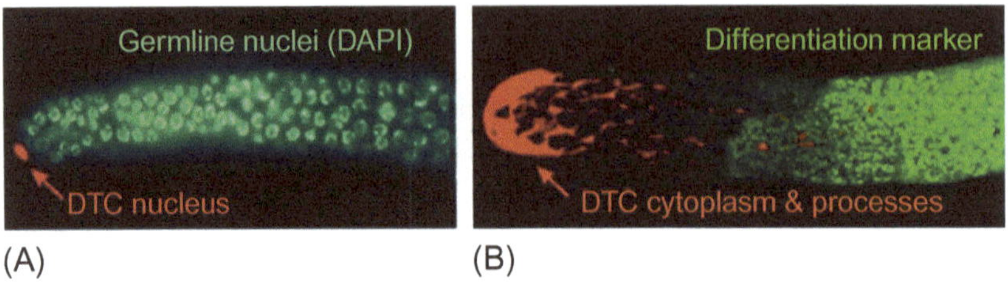

FIG. 3.2 The distal tip cell (DTC) in *C. elegans*. (A) The DTC nucleus (red) is localized at the distal end of each gonadal arm. Germ cell nuclei visualized using DAPI (cyan). (B) The DTC (red), highlighted with cytoplasmic green fluorescent protein, shows the cell body at the distal end and DTC processes extending proximally. The onset of germline differentiation is observed using the GLD-1 marker (green). *Modified from Kimble and Seidel, 2013, StemBook under the terms of the Creative Commons Attribution License.*

a meiotic state, resulting in the production of haploid gametes (Scadden, 2014; Spradling et al., 2011). In 1981 Kimble and White showed by ablation that the DTC was essential for sustaining the GSCs (Kimble and White, 1981). This simple niche requires no other cell types to support it.

Signaling Differentiation—Delta/Notch Pathway

The mechanism that underlies DTC control of GSC fate is Delta/Notch signaling (Fig. 3.3). Delta, the Notch receptor ligand that was introduced in Chapter 2, is a member of the Delta/Serrate/LAG-2 (DSL) protein family. The DTC expresses DSL member LAG-2 (Henderson et al., 1994; Tax et al., 1994). Notch receptors are transmembrane proteins with an intracellular and extracellular domain. The GSCs express the Notch-type receptor GLP-1 (Crittenden et al., 1994). When LAG-2 contacts GLP-1, the intracellular domain of GLP-1 is cleaved and then translocates to the

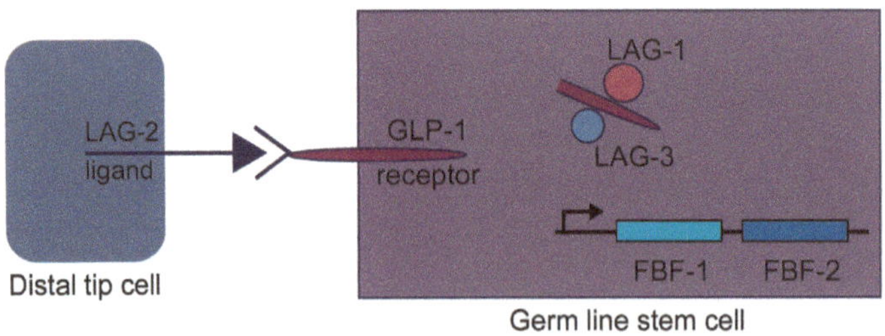

FIG. 3.3 Delta/Notch signaling represses differentiation in the germline stem cells. The Delta/Serrate/LAG-2 (DSL) member LAG-2 on the distal tip cell (DTC) activates the Notch-type receptor GLP-1 resulting in receptor cleavage. The GLP-1 intracellular domain is translocated to the nucleus and forms a complex with a DNA binding protein LAG-1, and transcriptional co-activator LAG-3 that results in the expression of two RNA-binding proteins, FBF-1 and FBF-2. The repression of differentiation genes by FBF-1 and FBF-2 ensures self-renewal of the GSCs.

nucleus where it forms a complex with the DNA binding protein LAG-1 and the transcriptional co-activator LAG-3 (Petcherski and Kimble, 2000). The complex of GLP-1, LAG-1, and LAG-3 activates the transcription of two RNA-binding proteins, FBF-1 and FBF-2 (Zhang et al., 1997). FBF proteins bind to RNA in a sequence-specific manner that suppress protein translation. In this case, a regulatory element for a whole host of genes responsible for differentiation is suppressed by FBF, which blocks GSC differentiation, thereby favoring self-renewal. Thus the DTC is a single-cell niche that directly regulates GSCs.

Fast-Cycling Niche: Intestinal Stem Cell Niche of Drosophila

The *Drosophila* midgut, introduced in Chapter 2, consists of an epithelial layer of two types of columnar cells—nutrient absorbing enterocytes (ECs) and digestive hormone secretory enteroendocrine cells (EEs). Intestinal stem cells (ISCs) at the base of the epithelia give rise to both EC and EE cells (Fig. 2.4B). Regulation of intestinal stem cell division is influenced by several different factors. Homeostasis requires a constant production of new cells, but may be upregulated 10–100 fold in response to inflammation or injury (Jiang and Edgar, 2011; Jiang et al., 2009; Chapter 2, Fig. 2.4). Scientists have experimented with several types of injury or infection in *Drosophila* gut including expression of the cell death gene *reaper* in ECs, providing food infused with cytotoxic agents, or by infecting the gut with bacteria (Amcheslavsky et al., 2009; Apidianakis et al., 2009; Biteau et al., 2011; Buchon et al., 2013; Cronin et al., 2009; Jiang et al., 2009). The cell signaling between damaged ECs and ISCs demonstrates a niche interaction that affects ISC proliferation.

Signaling proliferation—JAK/STAT Pathway

Damaged or dying ECs produce *cytokines* that influence ISC proliferation by upregulating the JAK/STAT pathway (Van Deuren et al., 1992). JAK/STAT signaling is a well-characterized pathway that regulates gene expression and can be activated by growth factors or hormones (Fig. 3.4; (Hombría and Brown, 2002; Nicolas et al., 2013). In the *Drosophila* gut, upregulation of JAK/STAT signaling increases ISC mitosis, ensuring a regenerative response to the injured epithelium (Jiang et al., 2009). The JAK/STAT signaling pathway involves several players: the interleukin 6 (IL-6)-like cytokine ligand called Unpaired (Upd) that is secreted by the ECs; the receptor Domeless (Dome) expressed on the ISCs; a Janus *kinase* (JAK) called Hopscotch; and a member of the signal transducers and activators of transcription (STAT) family STAT92E (Arbouzova, 2006; Brown et al., 2001). The binding of Upd to Dome activates Hopscotch. Hopscotch then phosphorylates Dome, providing a docking site for STAT92E (Rawlings,

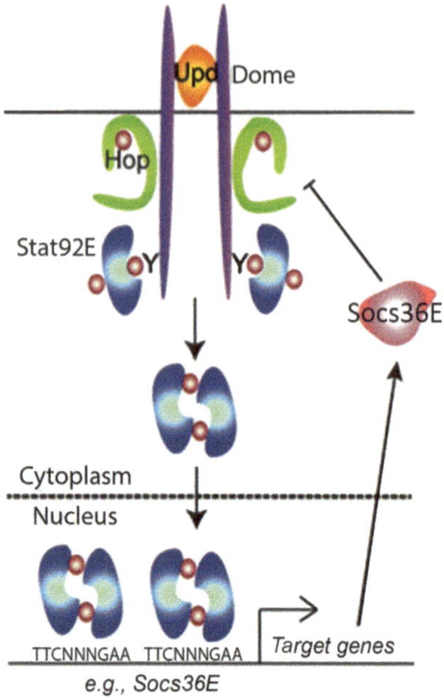

FIG. 3.4 The JAK/STAT signaling pathway. The Domeless (Dome) receptor (purple) is activated when it is bound by its ligand, Unpaired (Upd; orange). Dome activates the Janus kinase called Hopscotch (green), which phosphorylates (indicated by red circles) Dome. STAT92E (blue) can dock on the phosphorylated site of the Dome receptor, form a phosphorylated dimer and translocate to the nucleus. There, the STAT92E dimer targets genes that affect proliferation as well as the Socs36E gene that negatively regulates JAK–STAT signaling. *Reprinted with permission from Amoyel, M., Bach, E.A., 2012. Functions of the Drosophila JAK-STAT pathway: lessons from stem cells. JAK-STAT 1, 176–183.*

2004). Once STAT92E docks on Dome, dimers of STAT92E form and are phosphorylated before translocating to the nucleus. Once in the nucleus, STAT92E dimers cause the expression of growth factors that affect the rate of stem cell division (Zhou et al., 2013). Another target gene of STAT92E is Socs36E, a negative regulator of JAK/STAT signaling (Buchon et al., 2009). Negative regulation is important for turning off signaling. A recent study found evidence that JAK/STAT signaling by visceral muscle also influences ISC proliferation (Zhou et al., 2013). Thus, the *Drosophila* midgut niche includes a supportive visceral muscle base and regulatory epithelial cells of the midgut that influence ISC proliferation.

Slow-Cycling Niche: Mammalian Hair Follicle Stem Cell Niche

Compared to *Drosophila* ISCs, hair follicle stem cells (HFSCs) are on the other end of the proliferation spectrum. HFSCs are maintained during extended periods of *quiescence* (Lay et al., 2016). Quiescence is a state in which stem cell integrity is maintained without cell division. Quiescence protects stem cells over a lifetime from telomere erosion, replicative errors, and DNA damage due to stress and normal aging (Arai and Suda, 2008).

Hair production is accomplished by an organ found beneath the skin called the hair follicle that consists of an upper permanent region and a lower cycling region (Fig. 3.5). The base of the lower hair follicle or end bulb consists of several layers including the hair shaft, inner root sheath and outer root sheath from inside to outside, respectively (Fig. 3.5A;

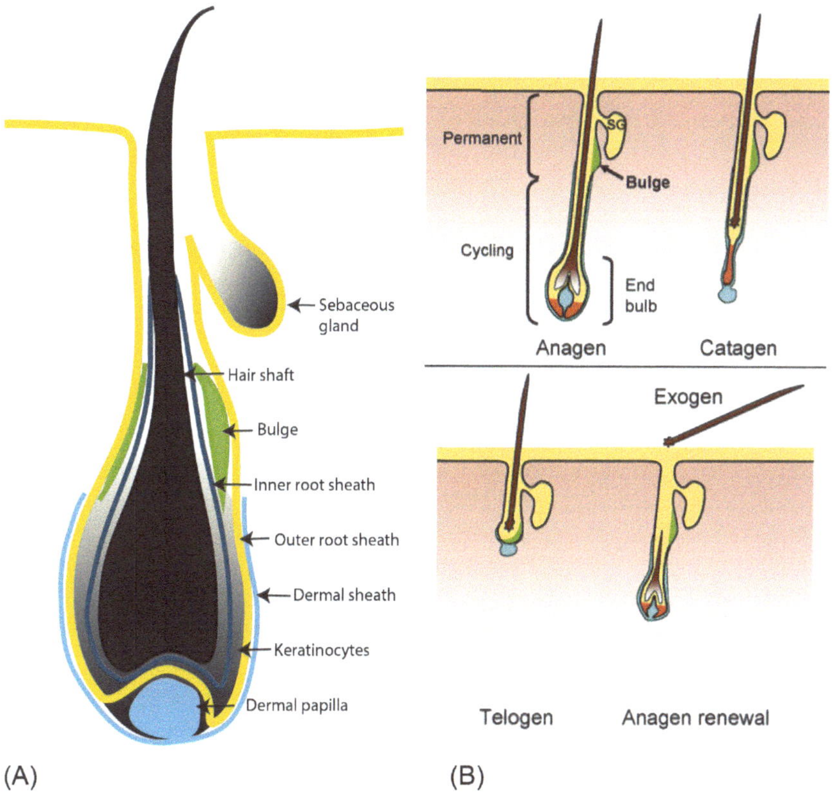

FIG. 3.5 The cyclical phases of the hair follicle. (A) The cellular layers of the hair follicle. During anagen (growth phase), the bulge (green), where hair follicle stem cells (HFSCs) reside is separated from the dermal papilla (blue) by the growing lower hair follicle. (B) Anagen and catagen (regression phase) occur over a period of 3 days. The quiescent telogen phase may last up to 4 weeks. HFSCs in the bulge self-renew during anagen, giving rise to keratinocytes that remodel the lower hair follicle (red) and produce hair shaft growth. During catagen the lower hair follicle recedes, bringing the dermal papilla in contact with the bulge. BMP4 secretion from the dermal papilla maintains the hair follicle in telogen. The hair shaft may be shed during the exogen phase after which the hair follicle cycles back to anagen. *(B) Modified with permission from Waters, J.M., Richardson, G.D., Jahoda, C.A.B., 2007. Hair follicle stem cells. Semin. Cell Dev. Biol. 18, 245–254.*

Margadant et al., 2010). The inner and outer root sheath layers extend toward the epidermis to surround the hair shaft. Below the bulb is the dermal papilla surrounded by the dermal sheath, a layer of cells that extends to surround the outer root sheath (Yang and Cotsarelis, 2010). Protruding from the outer root sheath in the upper hair follicle is the sebaceous gland that produces the lipids that make skin waterproof (Smith and Thiboutot, 2008). Just below the sebaceous gland is the bulge, a fitting description of its anatomy as it swells from the outer root sheath and contains the HFSCs.

The HFSC niche consists of several cell types, including fibroblasts of the surrounding dermis, subcutaneous adipocytes, the *keratinocyte* cell layer within the bulge as well as the dermal papilla at the base of the bulb (Solanas and Benitah, 2013). Fibroblasts, adipocytes and keratinocytes all have an ectodermal origin, while the dermal papilla is derived from the mesenchyme (Schneider et al., 2009). Signaling crosstalk between the niche cells orchestrates the constant growth and regeneration of hair. HFSCs can give rise to a new hair follicle, the epidermis and the sebaceous gland (Schneider et al., 2009).

The HFSCs cycle through three phases that remodel the end bulb: anagen, catagen, and telogen (Fig. 3.5B). In mice, it takes 3 days to cycle through anagen, catagen and telogen after which quiescence may be sustained for up to 4 weeks (Greco et al., 2009; Van Mater et al., 2003).

Anagen (growing phase) is a regenerative period when HFSCs self-renew. As the hair fiber lengthens and inner root sheath expands toward the surface of the skin, the outer root sheath undergoes down-growth deeper below the skin around the elongating hair follicle and dermal papilla (Rompolas et al., 2013; Waters et al., 2007). Concentric rings of keratinocytes make up the base of the end bulb called the matrix (Gat et al., 1998). HFSCs give rise to the keratinocytes, which are transit-amplifying progenitors that produce hair shaft growth (Cotsarelis et al., 1990). HFSCs return to a quiescent state midway through the anagen phase while keratinocytes continue to remodel the hair follicle.

Once a mature hair follicle is established, catagen (regression phase) is initiated in which the epithelial keratinocytes of the end bulb undergo apoptosis. Catagen leads to regression of the lower hair follicle in which the cells surrounding the dermal papilla recede. Meanwhile, the dermal papilla contracts upward bringing it into contact with the bulge (Gat et al., 1998). In the upper hair follicle the bulge and gland are stable structures that remain as well as the hair shaft. Catagen is followed by the quiescent telogen period (Cotsarelis et al., 1990; Tiede et al., 2007).

Signaling Quiescence—BMP and Wnt Pathway

During telogen (resting phase), cells of the hair follicle niche produce signals that maintain quiescence. Epidermal fibroblasts surrounding the bulge and subcutaneous adipocytes underlying the base of the bulb secrete members of the bone morphogenetic protein (BMP) family, signaling molecules that maintain quiescence (Lay et al., 2016). BMPs mediate signaling by binding BMP receptors on the HFSCs that activate the SMAD or MAP kinase pathway via the phosphorylation of SMAD proteins or MAP kinases (Fig. 3.6). The phosphorylated SMAD proteins are translocated to the nucleus resulting in gene expression changes that suppress proliferation. The phosphorylated MAP kinase translocates to the nucleus and activates transcription factors that regulate entry into catagen, and suppress proliferation (Akilli Öztürk et al., 2015). Under these circumstances, the activation of SMAD and MAP kinase signaling have complementary roles.

The dermal papilla is in close proximity to the bulge during telogen where it secretes BMP4. Keratinocytes that reside within the bulge secrete BMP6 and fibroblast growth factor (FGF) 18 that inhibit proliferation (Hsu et al., 2011). FGFs bind FGF receptors that can activate the MAP kinase pathway. One of the cytoplasmic functions of *unphosphorylated* MAP kinase is inhibitory phosphorylation of SMAD1, counteracting BMP signaling. Thus activation by FGF of the MAP kinase pathway drives the cell toward quiescence.

Anagen reentry relies on reciprocal Wnt signaling between the keratinocytes and the dermal papilla. Wnt family members are ligands that bind to Frizzled (Fz) receptors. Wnts can activate several different signaling pathways, but the most well known is the canonical Wnt/β catenin pathway (Fig. 3.7). Fz receptors activate a protein called Disheveled (Dsh) which prevents a complex of proteins, including glycogen synthase kinase 3 (GSK-3), from phosphorylating and proteolyzing (breaking down) molecules of β catenin. Inhibition of the GSK-3 complex allows β catenin to enter the nucleus, where it displaces a repressor protein and allows the transcription of specific target genes in the hair follicle, genes that initiate anagen.

Wnt signaling causes a change in secreted factors from the dermal papilla that act on adjacent keratinocytes (Chi et al., 2013; Enshell-Seijffers et al., 2010). In keratinocytes, Wnt signaling fuels the initial stages of anagen (Enshell-Seijffers et al., 2010; Solanas and Benitah, 2013). In response, the dermal papilla then begins to secrete BMP inhibitors that allow HFSCs to respond to activating cues of proliferation, such as β catenin signaling, that maintain proliferation through anagen (Solanas and Benitah, 2013). HFSCs express a protein called Tbx-1 that appears to act as a sensor of BMP signaling and may be a key player in balancing quiescence versus proliferation (Chen et al., 2012). Generation and growth of the cycling portion of the hair follicle, including the outer root sheath, distances the dermal papilla from the bulge and the HFSC-proliferating signals. Catagen appears to be triggered when the dermal papilla decreases production of β catenin, thereby decreasing the transcription of proliferation genes (Enshell-Seijffers et al., 2010). Thus, the niche of the HFSCs is a complex interaction of several different cell types that contribute to the control of hair follicle cycling.

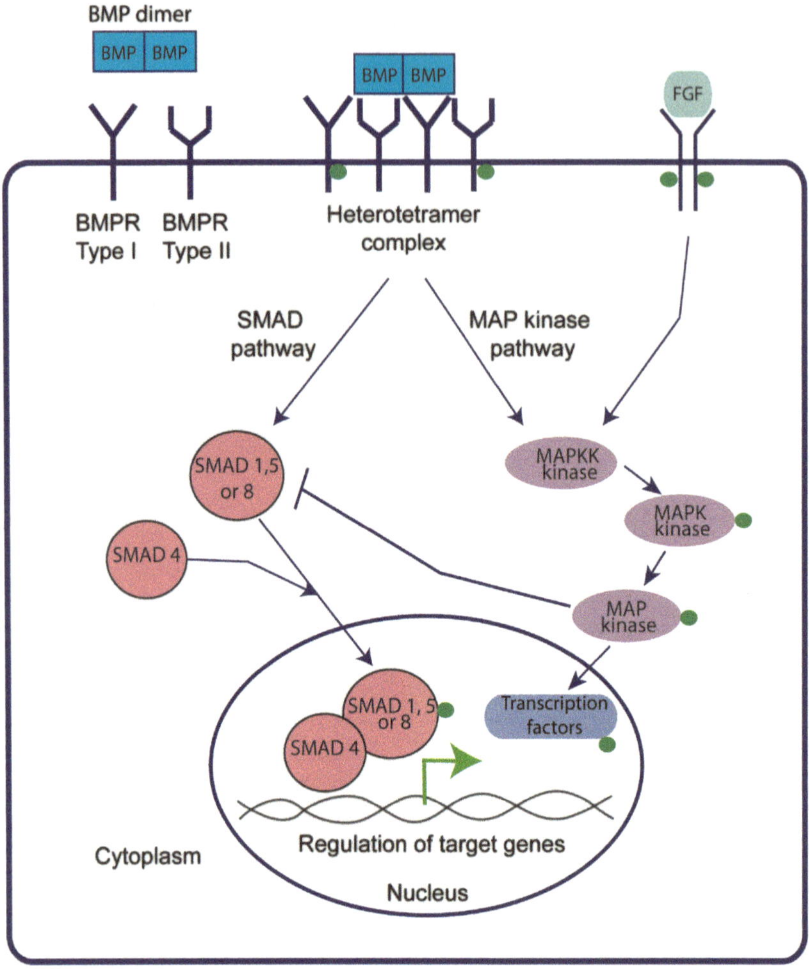

FIG. 3.6 BMP and FGF signaling pathways. The two BMP receptor subtypes form a heterotetramer when a BMP dimer binds to Type II receptors causing the phosphorylation of the Type I BMP receptor. SMAD proteins are signal transducers and transcriptional modifiers. When the Type I BMP receptor is phosphorylated it leads to the phosphorylation of SMADs 1, 5, or 8, which then combines with SMAD4 (pink circles) and translocates to the nucleus. There, the SMAD complex regulates the transcription of target genes involved in proliferation. Type I BMP receptor phosphorylation also causes a sequential phosphorylation of three kinase molecules starting with mitogen-activated kinase kinase (MAPKK), followed by MAPK kinase, and finally MAP kinase, the last of which translocates to the nucleus where it phosphorylates and activates transcription factors that regulate specific target genes. In the cytoplasm, the unphosphorylated form of MAPK can phosphorylate SMAD1 at an alternative site that inhibits SMAD signaling. FGF receptors dimerize when bound by an FGF molecule resulting in autophosphorylation and activation of the MAP kinase pathway.

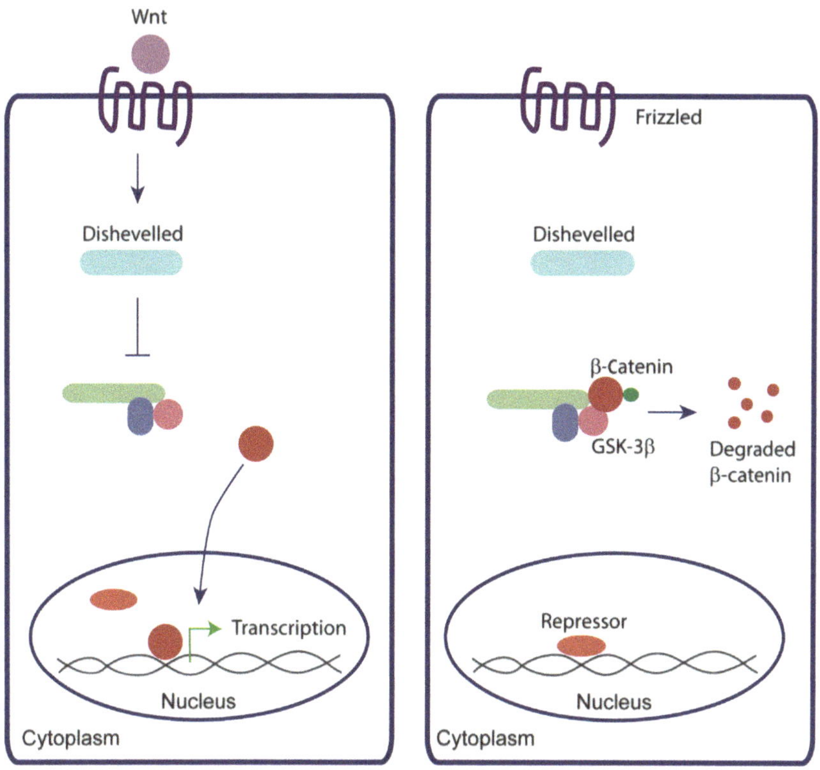

FIG. 3.7 Wnt signal activation of the canonical β-catenin pathway. When Wnt binds to a Frizzled receptor, the receptor activates the intracellular protein Disheveled. Activated Disheveled inhibits a complex of proteins that includes glycogen synthase kinase 3 (GSK-3b), which allows β-catenin to enter the nucleus where it displaces a protein that represses transcription. In the absence of Wnt, β-catenin is phosphorylated by GSK-3b resulting in its degradation.

NICHES WITHIN A NICHE: MAMMALIAN HEMATOPOIETIC STEM CELL NICHE

On a daily basis the average human body's production of 200 billion red blood cells is accomplished by a large population of stem cells. Hematopoiesis is the production of all blood cells, including red blood cells that carry oxygen to all areas of the body, white blood cells (lymphocytes and myelocytes) that fight infections and disease and platelets that mediate blood clotting and wound repair. Thirty years after they were identified in the hematopoietic system by Till and McCulloch, the ability of hematopoietic stem cells (HSCs) to give rise to all blood cell types was demonstrated in vitro (Becker et al., 1963; Spangrude et al., 1988; Till et al., 1964).

Hematopoiesis involves the production of a hierarchy of progenitor cells (see Chapter 1, Fig. 1.4). Long-term HSCs are largely quiescent to maintain the stem cell population. The short-term HSCs contribute to the production of the nine blood cell types. There are three levels of progenitor cells that have decreasing potential as they become more closely related to the final committed cells. A HSC first gives rise to two different multipotent progenitors, the common lymphocyte progenitor (CLP) and common myelocyte progenitor (CMP) called primitive progenitor cells. CMPs and CLPs each produce two committed precursor cells. Finally, the lineage-committed cells arise from the committed precursor cells to produce the nine differentiated cell types.

HSC behaviors include self-renewal of short-term HSCs, retention/maintenance of long-term HSCs, differentiation and quiescence. HSCs may be unique in that they can *mobilize* at a given signal and exit the niche into the blood stream. Conversely, HSCs return due to *homing* signals and *engraft* back into the niche. HSC behavior is influenced by its current genetic state (gene expression patterns), stress or injury, proximity to specific cellular features in the niche and secretion of factors by its progeny or other cell types within the niche.

Description of Cells/Anatomy That Make up the HSC Niche

The HSC niche is found within the spongy bone marrow called the medullary cavity, which contains a large number of cell types that support and influence hematopoiesis (Fig. 3.8). The inner surface of the bony tissue surrounding the medullary cavity is called the endosteum, a type of connective tissue primarily made up of osteoblast cells interspersed with osteoclasts. Osteoblasts contribute to the extracellular matrix production of the bone, while osteoclasts absorb bone and other endosteal components (Nagasawa et al., 2011). The endosteum and cells of the medullary cavity are derived from the mesenchyme, including stromal cells (connective tissue cells), MSCs, *reticular cells*, and adipocytes (Cordeiro-Spinetti et al., 2015; Klimczak and Kozlowska, 2016; Taichman, 2005). The reticular cells of the perivascular and nonperivascular type reside close to the vasculature or osteoblasts, respectively, and have long processes, creating physical support and cell-cell contacts (Bianco et al., 2001; Cordeiro-Spinetti et al., 2015). Sympathetic neurons are also present in the medullary cavity and are myelinated by Schwann cells or ensheathed by nonmyelinating Schwann cells that regulate bone formation and osteoblast function (Jessen and Mirsky, 2005; Takeda et al., 2002).

The endosteum and medullary cavity are highly vascularized areas containing two types of blood vessels, arterioles and sinusoids (Kunisaki et al., 2013). Arterioles are composed of elastic laminae and smooth muscle cells that can contract to increase blood pressure. The sinusoids are

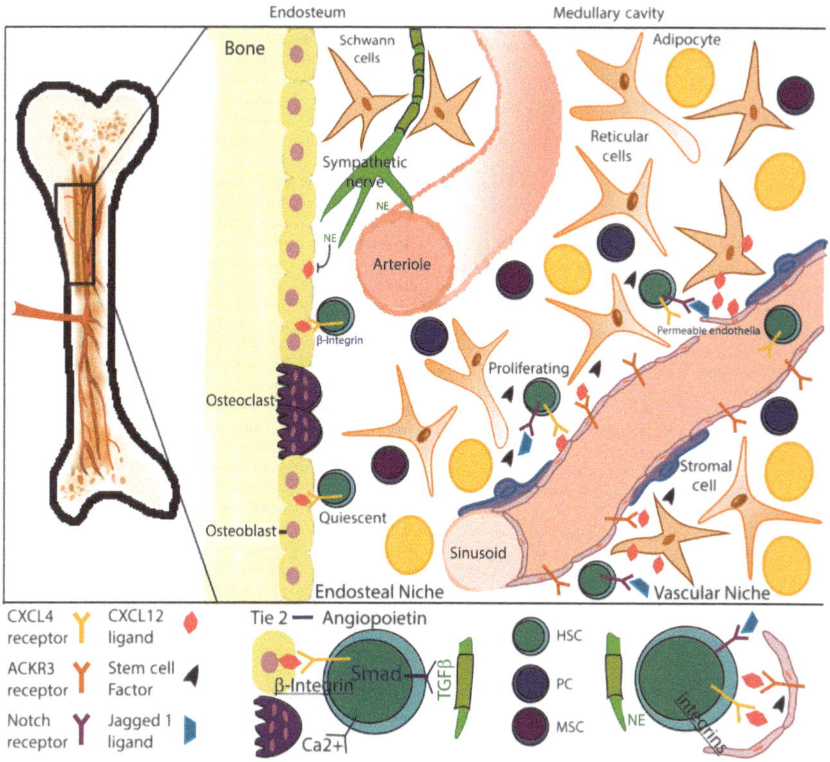

FIG. 3.8 The HSC niche can be divided into endosteal and vascular niches within the medullary cavity. The endosteum consists of osteoclasts and osteoblasts. The medullary cavity contains a number of cell types that contribute to the HSC niche, including the arterioles, epithelial cells of the sinusoids, reticular cells, stromal cells, adipocytes, and sympathetic nerves myelinated, or ensheathed, by Schwann cells. The hypoxic endosteal niche attracts CXCR4-expressing HSCs to osteoblasts expressing CXCL12. Osteoblasts maintain HSC quiescence by β-integrin adherence and Tie2 interactions with Angiopoietin on HSCs. As osteoclasts resorb bone, high calcium concentrations activate receptors on HSCs. *From Boulais, P.E., Frenette, P.S., 2015. Making sense of hematopoietic stem cell niches. Blood 125, 2621–2629.*

capillaries made up of endothelial cells that have large gaps between them, and are highly permeable to blood, water-soluble proteins, hormones and HSCs.

The microenvironment of the HSC niche can be divided into an endosteal (or osteoblastic) and vascular niche (Fig. 3.8). The arterioles and osteoblasts are associated preferentially with long-term HSCs and multipotent progenitors maintained in a quiescent state (Arai and Suda, 2008; Boulais and Frenette, 2015; Kubota et al., 2008). The endosteal niche is hypoxic, achieved by its distance from sinusoids. The sinusoids host the vascular niche, which is associated with proliferative HSCs that sustain blood cell production (Arai and Suda, 2008; Boulais and Frenette, 2015;

Kunisaki et al., 2013). The vascular niche is oxygen rich to support differentiated cells and hematopoiesis by HSCs and lineage-committed progenitors (Yin and Li, 2006).

HSCs can migrate between bone marrow compartments (e.g., femur to tibia), the spleen, or liver (Morrison and Spradling, 2008). Mobilization out of the niche can be induced chemically or by stress, resulting in HSCs proliferating where blood cells are needed the most (Pawig et al., 2015). For example, studies have shown that liver damage due to chronic alcoholism, acetaminophen overexposure, or viral hepatitis can be repaired by HSCs (Mark et al., 2010). Colony stimulating factors (CSF) were discovered in 1966 and granulocyte (G)-CSF, expressed in vivo by the sinusoidal endothelial cells, can rapidly mobilize blood cell progenitors (Bradley and Metcalf, 1966; Chatta et al., 1994; Ichikawa et al., 1966). Patients with compromised liver function are administered (G)-CSF and their HSCs are collected and expanded in vitro, after which the HSCs are injected back into the hepatic artery (Lorenzini et al., 2008). In a small study of patients suffering from alcoholic liver cirrhosis, the infusion of their HSCs after collection and expansion improved the outcome of five of nine patients (Pai et al., 2008). Mobilizing HSCs for collection has begun to replace the painfully invasive procedure of surgically harvesting bone marrow from donors (Khurana et al., 2014). The collected HSCs can be transplanted to other patients who have blood disorders, a compromised immune system or following chemotherapy (Box 3.1). Some patients have a reduced response to G-CSF due to chemotherapy or radiation, further emphasizing the importance of understanding the mechanisms that regulate HSC mobilization (Khurana et al., 2014).

BOX 3.1

SPOTLIGHT ON ACUTE MYELOID LEUKEMIA

Acute myeloid leukemia (AML) is a cancer of the bone marrow most commonly occurring in patients over 40 years old. AML occurs when a mutation to one of the blood cell precursors is unable to differentiate, blocking normal hematopoiesis (Chen et al., 2010; Fontemaggi et al., 2015). The resulting phenotype is the production of an abnormal committed precursor cell, namely the granulocytes and macrophage progenitor, or myeloblast that would normally give rise to red blood cells, platelets and some types of white blood cells (see Fig. 1.4). There are several different subtypes of AML, described using two different descriptor scales, however, one of the main determinants is which committed precursor cell has become cancerous (http://www.cancer.net/cancer-types/leukemia-acute-myeloid-aml/subtypes). Several mechanisms can contribute

BOX 3.1

SPOTLIGHT ON ACUTE MYELOID LEUKEMIA—CONT'D

to AML including transcriptional dysregulation, epigenetic modifications such as alterations in DNA methylation that alter implicated genes (activation/repression), and *microRNA* (miRNA) that can act as either oncogenes or tumor suppressors in AML (Chen et al., 2010).

There are four standard treatments for AML: chemotherapy (the primary treatment), radiation therapy, other drug therapy and stem cell transplant. Chemotherapy and radiation therapy are both used to kill the abnormal cells or to prevent them from dividing. Depending on the subtype of AML, 65% of patients may remain disease-free after chemotherapy alone.

HSC transplant is the first option for patients diagnosed with poor-risk leukemia, in which chemotherapy only has a 20% chance of keeping the patient disease-free. There are two types of HSC treatment: allogenic or autologous stem cell transplant. An autologous transplant involves the collection and freezing of the AML patient's own stem cells followed by chemotherapy treatment. The frozen stem cells are then thawed and infused back into the patient's blood stream. An allogenic transplant is when HSCs are obtained from the blood or bone marrow of another person with matching human leukocyte antigens (HLA), the proteins expressed on the white blood cells. ED Thomas (Chapter 1) published the first study in 1977 showing that allogeneic marrow transplant following chemotherapy and irradiation was effective to treat acute leukemia (Thomas et al., 1977). Approximately 9000 allogeneic HSC transplants were performed in 2010 (Magenau and Couriel, 2013).

Several changes are observed in the leukemic stem cell niche depending on the AML subtype. Some of the changes may include loss of sympathetic fibers and Schwann cells, damage to the vasculature and increased *angiogenesis* and loss of normal bone (Sánchez-Aguilera and Méndez-Ferrer, 2017). The interactive nature of the HSC niche is such that any loss to nerve, vascular or bone in the medullary cavity can cause loss of MSCs, or abnormal MSCs production, abnormal differentiation of mesenchymal stromal cells and/or osteoprogenitors and alterations in HSC maintenance. For example, leukemic myeloid cells (abnormal precursor cells) can stimulate multipotent stromal cells in the endosteal niche resulting in an overproduction of osteoblastic cells and bone thickening that prevents HSC maintenance (Schepers et al., 2013). Conversely, AML can cause osteoblastic inhibition resulting in decreased bone formation, again resulting in a disruption of HSC maintenance (Frisch et al., 2012). A better understanding of the cellular interplay within the HSC niche has led to insights that may result in improved therapies to treat AML and other leukemia subtypes.

SIGNALING MOBILIZATION, HOMING, AND QUIESCENCE—CXCL12/CXCR4 PATHWAY

Chemokines are proteins that regulate the mobilization or migration of cells. Chemokines bind receptors on affected cells and bind to nonreceptor sites at fixed locations to create a chemokine gradient that attracts migrating cells (Barinov et al., 2017). C-X-C motif chemokine 12 (CXCL12), also known as stromal cell-derived factor 1 (SDF1), is a chemoattractant homing cue to HSCs expressing the G protein-coupled receptor CXCR4. CXCL12 is a versatile ligand in that it can bind to CXCR4 monomers, CXCR4 homodimers, ACKR3 (aka CXCL7) receptors and CXCR4/ACKR3 heterodimers. Depending on the type of receptor or dimer combination it binds, a different signal transduction pathway is activated. In two known instances the binding results in receptor degradation (Pawig et al., 2015). Several populations in the bone marrow niche regulate CXCR4-expressing HSC behaviors by CXCL12 expression.

In the vascular niche, mobilization and self-renewal are regulated by several cell types. Sinusoid endothelial cells and perivascular reticular cells express CXCL12 that attracts HSCs to the bone marrow as part of the homing signaling process. The sinusoid endothelial cells also express high levels of ACKR3 that bind excess CXCL12 molecules, which may serve to either prevent desensitization of CXCR4 or modify the gradient of CXCL12 for migrating cells to follow (Karpova and Bonig, 2015). CXCL12 causes an upregulation of several integrins in CXCR4-expressing HSCs that enable engraftment to the sinusoids (Pawig et al., 2015; Peled et al., 1999, 2000). In zebrafish, sinusoid endothelial cells can dynamically *remodel* their morphology in order to "cuddle" HSCs that have just exited the blood stream (Tamplin et al., 2015). The remodeling serves to lodge HSCs on the outer wall of the endothelium and orient the plane of division. Upon engraftment Notch receptor ligands such as Jagged-1 expressed by sinusoid endothelial cells promote HSC self-renewal and proliferation (Butler et al., 2010; Poulos et al., 2013). Further, sinusoidal endothelial cells and perivascular reticular cells express stem cell factor, a cytokine that promotes HSC maintenance (Ding et al., 2012). Finally, the sympathetic nervous system also regulates HSC behavior in a circadian manner (Méndez-Ferrer et al., 2008). The release of norepinephrine, peaking five hours into the daylight cycle, mobilizes HSCs, while decreasing osteoblast production of CXCL12 (Boulais and Frenette, 2015; Katayama et al., 2006).

In the endosteal niche, the hypoxic environment upregulates CXCL12 expression in osteoblasts, which attracts CXCR4-expressing HSCs from the vascular niche (Khurana et al., 2014; Pawig et al., 2015). Once HSCs are near the endosteum, several factors maintain HSCs in a quiescent state. The cell surface receptor *β-integrin* mediates HSC adherence (localization), retention and contact-dependent inhibition of proliferation

(Arai and Suda, 2008; Lo Celso and Scadden, 2011; Sugiyama et al., 2006). The tyrosine receptor kinase Tie 2 expressed by the osteoblasts is a ligand for Angiopoietin 1, which is expressed by the HSCs. Tie2/Angiopoietin interactions further adhere HSCs to the endosteal niche (Arai et al., 2004). Maintenance and localization is promoted by high calcium concentrations that activate calcium receptors on HSCs when osteoclasts release calcium after bone resorption (Adams et al., 2006). Finally, HSC quiescence is maintained by the nonmyelinating Schwann cells by activation of SMAD signaling, although it is through activation of *TGFβ receptors* instead of BMP receptors (Yamazaki et al., 2011). Thus there is a balance between quiescence and proliferation based on CXCR4 attraction to either CXCL12 in the osteoblastic niche or CXCL12 in the vascular niche (Karpova and Bonig, 2015).

Our understanding of the HSC niche is still a work in progress. Its lack of discrete compartmentalization adds a layer of difficulty to determine the role of specific cell populations and the factors they secrete. Cancers such as leukemia, blood and bone disorders, and immunodeficiency all have a connection to an altered niche. Scientists and clinicians alike are driven to understand this complex niche and apply that knowledge to curing disease.

SUMMARY

This chapter introduces four types of stem cell niches that touch on the differences of niche composition and function. The description of each niche has been accompanied by a description of a signaling pathway that regulates certain aspects of stem cell behavior within that niche. Despite the wide variety of niches, the question remains as to whether the regulatory pathways are conserved for specific behaviors. For example, does the Delta/Notch signaling pathway regulate differentiation in all somatic stem cells as it does in the *C. elegans* DTC for GSCs? Does JAK/STAT signaling regulate proliferation in all somatic stem cells as it does in the *Drosophila* intestinal stem cell niche? These questions and more continue to be addressed in labs using models from *C. elegans* to fish and rodents. The relevance to human health could not be greater. For example, malignant tumors are caused by cells undergoing unchecked proliferation. A thorough understanding of cell proliferation can provide drug targets to prevent cancerous growth. Another strategy may be to induce quiescence in the cancer cells. Further, could we apply our understanding of homing to prevent cancer malignancy? One challenge is to target the drugs specifically to the cancer cells without harming somatic stem cells. Other areas of research in which stem cells are relevant are autoimmune and degenerative diseases, congenital disorders, burns and other injuries, to name a few. Research efforts worldwide are ongoing to unlock the promise of stem cells.

DEFINITIONS

Angiogenesis The production of new blood vessels.

β-Integrin A member of the integrin family, a large group of transmembrane receptors that adhere a cell's cytoskeleton to the extracellular matrix or other cells.

Chemokines Proteins arranged in a gradient on nonreceptor sites that attracts cells expressing receptors for the chemokine that induces the cells to mobilize or migrate.

Cytokines Hormone-like compounds secreted by a broad range of cells that have an effect on the cell of origin (autocrine), surrounding cells (paracrine) or a biological system (endocrine). A given cytokine can be produced by more than one cell type. For example, interleukin 6 (IL-6)-like cytokine was originally shown to play a role in B-cell maturation in the immune system; however, it is also involved in neurogenesis and mature neuron and glial responses (Erta et al., 2012; Roldán and Brieva, 1991).

Endosteum A type of connective tissue made up of osteoblasts, interspersed with osteoclasts that line the inner surface of the bony tissue surrounding the medullary cavity.

Engraft The process in which HSCs are recruited to the vascular HSC niche where they begin to proliferate.

Haploid A cell that has half the diploid number of chromosomes.

Homing When HSCs are induced by the chemokine CXCL12 to return from circulating in the peripheral blood to a hematopoietic niche through endothelial sinusoids.

Keratinocytes Cells of the epidermis that produce keratin.

Kinase An enzyme responsible for phosphorylation of a substrate through the transfer of a phosphate group from ATP.

MicroRNA A small, noncoding RNA fragment of 19–22 nucleotides that prevents expression of a protein by binding to the messenger RNA that would have produced it, resulting in the destruction of the double strand of RNA. MicroRNA is used by geneticists to suppress specific genes but has been found normally in cells playing different regulatory roles.

Mobilization A process that induces HSCs to enter the blood stream from the niche. Granulocyte-colony stimulating factor normally induces mobilization.

Quiescence A period in which stem cells do not divide; a state that sustains the integrity of stem cells.

Remodel To cause a cell to undergo morphological change.

Reticular cells A type of fibroblast with long processes that extend to other cell types or other reticular cells creating structural support within a tissue.

TGFβ receptors Transforming growth factor beta is a ligand that binds to any of four different splice versions of a transmembrane serine/threonine kinase (Trk) receptor.

References

Adams, G.B., Chabner, K.T., Alley, I.R., Olson, D.P., Szczepiorkowski, Z.M., Poznansky, M.C., Kos, C.H., Pollak, M.R., Brown, E.M., Scadden, D.T., 2006. Stem cell engraftment at the endosteal niche is specified by the calcium-sensing receptor. Nature 439, 599–603.

Akilli Öztürk, Ö., Pakula, H., Chmielowiec, J., Qi, J., Stein, S., Lan, L., Sasaki, Y., Rajewsky, K., Birchmeier, W., 2015. Gab1 and Mapk signaling are essential in the hair cycle and hair follicle stem cell quiescence. Cell Rep. 13, 561–572.

Amcheslavsky, A., Jiang, J., Ip, Y.T., 2009. Tissue damage-induced intestinal stem cell division in Drosophila. Cell Stem Cell 4, 49–61.

Apidianakis, Y., Pitsouli, C., Perrimon, N., Rahme, L., 2009. Synergy between bacterial infection and genetic predisposition in intestinal dysplasia. Proc. Natl. Acad. Sci. U. S. A. 106, 20883–20888.

Arai, F., Suda, T., 2008. Quiescent stem cells in the niche. In: StemBook. Harvard Stem Cell Institute, Cambridge, MA.

Arai, F., Hirao, A., Ohmura, M., Sato, H., Matsuoka, S., Takubo, K., Ito, K., Koh, G.Y., Suda, T., 2004. Tie2/angiopoietin-1 signaling regulates hematopoietic stem cell quiescence in the bone marrow niche. Cell 118, 149–161.

Arbouzova, N.I., 2006. JAK/STAT signalling in Drosophila: insights into conserved regulatory and cellular functions. Development 133, 2605–2616.

Barinov, A., Luo, L., Gasse, P., Meas-Yedid, V., Donnadieu, E., Arenzana-Seisdedos, F., Vieira, P., 2017. Essential role of immobilized chemokine CXCL12 in the regulation of the humoral immune response. Proc. Natl. Acad. Sci. 114, 2319–2324.

Becker, A.J., McCulloch, E.A., Till, J.E., 1963. Cytological demonstration of the clonal nature of spleen colonies derived from transplanted mouse marrow cells. Nature 197, 452–454.

Bianco, P., Riminucci, M., Gronthos, S., Robey, P.G., 2001. Bone marrow stromal stem cells: nature, biology, and potential applications. Stem Cells 19, 180–192.

Biteau, B., Hochmuth, C.E., Jasper, H., 2011. Maintaining tissue homeostasis: dynamic control of somatic stem cell activity. Cell Stem Cell 9, 402–411.

Boulais, P.E., Frenette, P.S., 2015. Making sense of hematopoietic stem cell niches. Blood 125, 2621–2629.

Bradley, T.R., Metcalf, D., 1966. The growth of mouse bone marrow cells in vitro. Aust. J. Exp. Biol. Med. Sci. 44, 287–299.

Brown, S., Hu, N., Hombr??a, J.C.-G., 2001. Identification of the first invertebrate interleukin JAK/STAT receptor, the Drosophila gene domeless. Curr. Biol. 11, 1700–1705.

Buchon, N., Broderick, N.A., Chakrabarti, S., Lemaitre, B., 2009. Invasive and indigenous microbiota impact intestinal stem cell activity through multiple pathways in Drosophila. Genes Dev. 23, 2333–2344.

Buchon, N., Osman, D., David, F.P.A., Yu Fang, H., Boquete, J.-P., Deplancke, B., Lemaitre, B., 2013. Morphological and molecular characterization of adult midgut compartmentalization in Drosophila. Cell Rep. 3, 1725–1738.

Butler, J.M., Nolan, D.J., Vertes, E.L., Varnum-Finney, B., Kobayashi, H., Hooper, A.T., Seandel, M., Shido, K., White, I.A., Kobayashi, M., et al., 2010. Endothelial cells are essential for the self-renewal and repopulation of notch-dependent hematopoietic stem cells. Cell Stem Cell 6, 251–264.

Chatta, G.S., Price, T.H., Allen, R.C., Dale, D.C., 1994. Effects of in vivo recombinant methionyl human granulocyte colony-stimulating factor on the neutrophil response and peripheral blood colony-forming cells in healthy young and elderly adult volunteers. Blood 84, 2923–2929.

Chen, J., Odenike, O., Rowley, J.D., 2010. Leukaemogenesis: more than mutant genes. Nat. Rev. Cancer 10, 23–36.

Chen, T., Heller, E., Beronja, S., Oshimori, N., Stokes, N., Fuchs, E., 2012. An RNA interference screen uncovers a new molecule in stem cell self-renewal and long-term regeneration. Nature 485, 104–108.

Chi, W., Wu, E., Morgan, B.A., 2013. Dermal papilla cell number specifies hair size, shape and cycling and its reduction causes follicular decline. Development 140, 1676–1683.

Cordeiro-Spinetti, E., Taichman, R.S., Balduino, A., 2015. The bone marrow endosteal niche: how far from the surface? J. Cell. Biochem. 116, 6–11.

Cotsarelis, G., Sun, T.T., Lavker, R.M., 1990. Label-retaining cells reside in the bulge area of pilosebaceous unit: implications for follicular stem cells, hair cycle, and skin carcinogenesis. Cell 61, 1329–1337.

Crittenden, S.L., Troemel, E.R., Evans, T.C., Kimble, J., 1994. GLP-1 is localized to the mitotic region of the *C. elegans* germ line. Development (Camb. Engl.) 120, 2901–2911.
Cronin, S.J.F., Nehme, N.T., Limmer, S., Liegeois, S., Pospisilik, J.A., Schramek, D., Leibbrandt, A., de Simoes, R.M., Gruber, S., Puc, U., et al., 2009. Genome-wide RNAi screen identifies genes involved in intestinal pathogenic bacterial infection. Science 325, 340–343.
Ding, L., Saunders, T.L., Enikolopov, G., Morrison, S.J., 2012. Endothelial and perivascular cells maintain haematopoietic stem cells. Nature 481, 457–462.
Enshell-Seijffers, D., Lindon, C., Kashiwagi, M., Morgan, B.A., 2010. Beta-catenin activity in the dermal papilla regulates morphogenesis and regeneration of hair. Dev. Cell 18, 633–642.
Erta, M., Quintana, A., Hidalgo, J., 2012. Interleukin-6, a major cytokine in the central nervous system. Int. J. Biol. Sci. 8, 1254–1266.
Fontemaggi, G., Bellissimo, T., Donzelli, S., Iosue, I., Benassi, B., Bellotti, G., Blandino, G., Fazi, F., 2015. Identification of post-transcriptional regulatory networks during myeloblast-to-monocyte differentiation transition. RNA Biol. 12, 690–700.
Frisch, B.J., Ashton, J.M., Xing, L., Becker, M.W., Jordan, C.T., Calvi, L.M., 2012. Functional inhibition of osteoblastic cells in an in vivo mouse model of myeloid leukemia. Blood 119, 540–550.
Gat, U., DasGupta, R., Degenstein, L., Fuchs, E., 1998. De novo hair follicle morphogenesis and hair tumors in mice expressing a truncated β-catenin in skin. Cell 95, 605–614.
Greco, V., Chen, T., Rendl, M., Schober, M., Pasolli, H.A., Stokes, N., Dela Cruz-Racelis, J., Fuchs, E., 2009. A two-step mechanism for stem cell activation during hair regeneration. Cell Stem Cell 4, 155–169.
Henderson, S.T., Gao, D., Lambie, E.J., Kimble, J., 1994. Lag-2 may encode a signaling ligand for the GLP-1 and LIN-12 receptors of *C. elegans*. Development (Camb. Engl.) 120, 2913–2924.
Hombría, J.C.-G., Brown, S., 2002. The fertile field of Drosophila JAK/STAT signalling. Curr. Biol. 12, R569–R575.
Hsu, Y.-C., Pasolli, H.A., Fuchs, E., 2011. Dynamics between stem cells, niche, and progeny in the hair follicle. Cell 144, 92–105.
Ichikawa, Y., Pluznik, D.H., Sachs, L., 1966. In vitro control of the development of macrophage and granulocyte colonies. Proc. Natl. Acad. Sci. U. S. A. 56, 488–495.
Jessen, K.R., Mirsky, R., 2005. The origin and development of glial cells in peripheral nerves. Nat. Rev. Neurosci. 6, 671–682.
Jiang, H., Edgar, B.A., 2011. Intestinal stem cells in the adult Drosophila midgut. Exp. Cell Res. 317, 2780–2788.
Jiang, H., Patel, P.H., Kohlmaier, A., Grenley, M.O., McEwen, D.G., Edgar, B.A., 2009. Cytokine/Jak/Stat signaling mediates regeneration and homeostasis in the Drosophila midgut. Cell 137, 1343–1355.
Karpova, D., Bonig, H., 2015. Concise review: CXCR4/CXCL12 signaling in immature hematopoiesis–lessons from pharmacological and genetic models. Stem Cells Dayt. Ohio 33, 2391–2399.
Katayama, Y., Battista, M., Kao, W.-M., Hidalgo, A., Peired, A.J., Thomas, S.A., Frenette, P.S., 2006. Signals from the sympathetic nervous system regulate hematopoietic stem cell egress from bone marrow. Cell 124, 407–421.
Khurana, S., Melacarne, A., Yadak, R., Schouteden, S., Notelaers, T., Pistoni, M., Maes, C., Verfaillie, C.M., 2014. SMAD signaling regulates CXCL12 expression in the bone marrow niche, affecting homing and mobilization of hematopoietic progenitors: regulation of CXCL12 expression in HSPC niche. Stem Cells 32, 3012–3022.
Kimble, J., 2014. *C. elegans* germline stem cells and their niche. In: StemBook.
Kimble, J.E., White, J.G., 1981. On the control of germ cell development in *Caenorhabditis elegans*. Dev. Biol. 81, 208–219.
Klimczak, A., Kozlowska, U., 2016. Mesenchymal stromal cells and tissue-specific progenitor cells: their role in tissue homeostasis. Stem Cells Int. 2016, 1–11.

Kubota, Y., Takubo, K., Suda, T., 2008. Bone marrow long label-retaining cells reside in the sinusoidal hypoxic niche. Biochem. Biophys. Res. Commun. 366, 335–339.

Kunisaki, Y., Bruns, I., Scheiermann, C., Ahmed, J., Pinho, S., Zhang, D., Mizoguchi, T., Wei, Q., Lucas, D., Ito, K., et al., 2013. Arteriolar niches maintain haematopoietic stem cell quiescence. Nature 502, 637–643.

Lay, K., Kume, T., Fuchs, E., 2016. FOXC1 maintains the hair follicle stem cell niche and governs stem cell quiescence to preserve long-term tissue-regenerating potential. Proc. Natl. Acad. Sci. 113, E1506–E1515.

Lo Celso, C., Scadden, D.T., 2011. The haematopoietic stem cell niche at a glance. J. Cell Sci. 124, 3529–3535.

Lorenzini, S., Isidori, A., Catani, L., Gramenzi, A., Talarico, S., Bonifazi, F., Giudice, V., Conte, R., Baccarani, M., Bernardi, M., et al., 2008. Stem cell mobilization and collection in patients with liver cirrhosis: G-CSF and stem cell mobilization in cirrhosis. Aliment. Pharmacol. Ther. 27, 932–939.

Magenau, J., Couriel, D.R., 2013. Hematopoietic stem cell transplantation for acute myeloid leukemia: to whom, when, and how. Curr. Oncol. Rep. 15, 436–444.

Margadant, C., Charafeddine, R.A., Sonnenberg, A., 2010. Unique and redundant functions of integrins in the epidermis. FASEB J. Off. Publ. Fed. Am. Soc. Exp. Biol. 24, 4133–4152.

Mark, A.L., Sun, Z., Warren, D.S., Lonze, B.E., Knabel, M.K., Williams, G.M.M., Locke, J.E., Montgomery, R.A., Cameron, A.M., 2010. Stem cell mobilization is life saving in an animal model of acute liver failure. Trans. Meet. Am. Surg. Assoc. 128, 192–198.

Méndez-Ferrer, S., Lucas, D., Battista, M., Frenette, P.S., 2008. Haematopoietic stem cell release is regulated by circadian oscillations. Nature 452, 442–447.

Morrison, S.J., Spradling, A.C., 2008. Stem cells and niches: mechanisms that promote stem cell maintenance throughout life. Cell 132, 598–611.

Nagasawa, T., Omatsu, Y., Sugiyama, T., 2011. Control of hematopoietic stem cells by the bone marrow stromal niche: the role of reticular cells. Trends Immunol. 32, 315–320.

Nicolas, C.S., Amici, M., Bortolotto, Z.A., Doherty, A., Csaba, Z., Fafouri, A., Dournaud, P., Gressens, P., Collingridge, G.L., Peineau, S., 2013. The role of JAK-STAT signaling within the CNS. JAK-STAT 2, e22925.

Pai, M., Zacharoulis, D., Milicevic, M.N., Helmy, S., Jiao, L.R., Levičar, N., Tait, P., Scott, M., Marley, S.B., Jestice, K., et al., 2008. Autologous infusion of expanded mobilized adult bone marrow-derived CD34+ cells into patients with alcoholic liver cirrhosis. Am. J. Gastroenterol. 103, 1952–1958.

Pawig, L., Klasen, C., Weber, C., Bernhagen, J., Noels, H., 2015. Diversity and inter-connections in the CXCR4 chemokine receptor/ligand family: molecular perspectives. Front. Immunol. 6.

Peled, A., Grabovsky, V., Habler, L., Sandbank, J., Arenzana-Seisdedos, F., Petit, I., Ben-Hur, H., Lapidot, T., Alon, R., 1999. The chemokine SDF-1 stimulates integrin-mediated arrest of CD34(+) cells on vascular endothelium under shear flow. J. Clin. Invest. 104, 1199–1211.

Peled, A., Kollet, O., Ponomaryov, T., Petit, I., Franitza, S., Grabovsky, V., Slav, M.M., Nagler, A., Lider, O., Alon, R., et al., 2000. The chemokine SDF-1 activates the integrins LFA-1, VLA-4, and VLA-5 on immature human CD34(+) cells: role in transendothelial/stromal migration and engraftment of NOD/SCID mice. Blood 95, 3289–3296.

Petcherski, A.G., Kimble, J., 2000. LAG-3 is a putative transcriptional activator in the *C. elegans* Notch pathway. Nature 405, 364–368.

Poulos, M.G., Guo, P., Kofler, N.M., Pinho, S., Gutkin, M.C., Tikhonova, A., Aifantis, I., Frenette, P.S., Kitajewski, J., Rafii, S., et al., 2013. Endothelial Jagged-1 is necessary for homeostatic and regenerative hematopoiesis. Cell Rep. 4, 1022–1034.

Rawlings, J.S., 2004. The JAK/STAT signaling pathway. J. Cell Sci. 117, 1281–1283.

Roldán, E., Brieva, J.A., 1991. Terminal differentiation of human bone marrow cells capable of spontaneous and high-rate immunoglobulin secretion: role of bone marrow stromal cells and interleukin 6. Eur. J. Immunol. 21, 2671–2677.

Rompolas, P., Mesa, K.R., Greco, V., 2013. Spatial organization within a niche as a determinant of stem-cell fate. Nature 502, 513–518.
Sánchez-Aguilera, A., Méndez-Ferrer, S., 2017. The hematopoietic stem-cell niche in health and leukemia. Cell. Mol. Life Sci. CMLS 74, 579–590.
Scadden, D.T., 2014. Nice neighborhood: emerging concepts of the stem cell niche. Cell 157, 41–50.
Schepers, K., Pietras, E.M., Reynaud, D., Flach, J., Binnewies, M., Garg, T., Wagers, A.J., Hsiao, E.C., Passegué, E., 2013. Myeloproliferative neoplasia remodels the endosteal bone marrow niche into a self-reinforcing leukemic niche. Cell Stem Cell 13, 285–299.
Schneider, M.R., Schmidt-Ullrich, R., Paus, R., 2009. The hair follicle as a dynamic miniorgan. Curr. Biol. 19, R132–R142.
Schofield, R., 1978. The relationship between the spleen colony-forming cell and the haemopoietic stem cell. Blood Cells 4, 7–25.
Smith, K.R., Thiboutot, D.M., 2008. Thematic review series: skin lipids. Sebaceous gland lipids: friend or foe? J. Lipid Res. 49, 271–281.
Solanas, G., Benitah, S.A., 2013. Regenerating the skin: a task for the heterogeneous stem cell pool and surrounding niche. Nat. Rev. Mol. Cell Biol. 14, 737–748.
Spangrude, G.J., Heimfeld, S., Weissman, I.L., 1988. Purification and characterization of mouse hematopoietic stem cells. Science 241, 58–62.
Spradling, A., Fuller, M.T., Braun, R.E., Yoshida, S., 2011. Germline Stem Cells. Cold Spring Harb. Perspect. Biol. 3, a002642.
Sugiyama, T., Kohara, H., Noda, M., Nagasawa, T., 2006. Maintenance of the hematopoietic stem cell pool by CXCL12-CXCR4 chemokine signaling in bone marrow stromal cell niches. Immunity 25, 977–988.
Sulston, J.E., Schierenberg, E., White, J.G., Thomson, J.N., 1983. The embryonic cell lineage of the nematode *Caenorhabditis elegans*. Dev. Biol. 100, 64–119.
Taichman, R.S., 2005. Blood and bone: two tissues whose fates are intertwined to create the hematopoietic stem-cell niche. Blood 105, 2631–2639.
Takeda, S., Elefteriou, F., Levasseur, R., Liu, X., Zhao, L., Parker, K.L., Armstrong, D., Ducy, P., Karsenty, G., 2002. Leptin regulates bone formation via the sympathetic nervous system. Cell 111, 305–317.
Tamplin, O.J., Durand, E.M., Carr, L.A., Childs, S.J., Hagedorn, E.J., Li, P., Yzaguirre, A.D., Speck, N.A., Zon, L.I., 2015. Hematopoietic stem cell arrival triggers dynamic remodeling of the perivascular niche. Cell 160, 241–252.
Tax, F.E., Yeargers, J.J., Thomas, J.H., 1994. Sequence of C. elegans lag-2 reveals a cell-signalling domain shared with Delta and Serrate of Drosophila. Nature 368, 150–154.
Thomas, E.D., Buckner, C.D., Banaji, M., Clift, R.A., Fefer, A., Flournoy, N., Goodell, B.W., Hickman, R.O., Lerner, K.G., Neiman, P.E., et al., 1977. One hundred patients with acute leukemia treated by chemotherapy, total body irradiation, and allogeneic marrow transplantation. Blood 49, 511–533.
Tiede, S., Kloepper, J.E., Bodò, E., Tiwari, S., Kruse, C., Paus, R., 2007. Hair follicle stem cells: walking the maze. Eur. J. Cell Biol. 86, 355–376.
Till, J.E., Mcculloch, E.A., Siminovitch, L., 1964. A stochastic model of stem cell proliferation, based on the growth of spleen colony-forming cells. Proc. Natl. Acad. Sci. U. S. A. 51, 29–36.
Van Deuren, M., Dofferhoff, A.S.M., Van Der Meer, J.W.M., 1992. Cytokines and the response to infection. J. Pathol. 168, 349–356.
Van Mater, D., Kolligs, F.T., Dlugosz, A.A., Fearon, E.R., 2003. Transient activation of beta-catenin signaling in cutaneous keratinocytes is sufficient to trigger the active growth phase of the hair cycle in mice. Genes Dev. 17, 1219–1224.
Waters, J.M., Richardson, G.D., Jahoda, C.A.B., 2007. Hair follicle stem cells. Semin. Cell Dev. Biol. 18, 245–254.
White, J.G., Southgate, E., Thomson, J.N., Brenner, S., 1986. The structure of the nervous system of the nematode Caenorhabditis elegans. Philos. Trans. R. Soc. Lond. Ser. B Biol. Sci. 314, 1–340.

Yamazaki, S., Ema, H., Karlsson, G., Yamaguchi, T., Miyoshi, H., Shioda, S., Taketo, M.M., Karlsson, S., Iwama, A., Nakauchi, H., 2011. Nonmyelinating Schwann cells maintain hematopoietic stem cell hibernation in the bone marrow niche. Cell 147, 1146–1158.
Yang, C.-C., Cotsarelis, G., 2010. Review of hair follicle dermal cells. J. Dermatol. Sci. 57, 2–11.
Yin, T., Li, L., 2006. The stem cell niches in bone. J. Clin. Invest. 116, 1195–1201.
Zhang, B., Gallegos, M., Puoti, A., Durkin, E., Fields, S., Kimble, J., Wickens, M.P., 1997. A conserved RNA-binding protein that regulates sexual fates in the *C. elegans* hermaphrodite germ line. Nature 390, 477–484.
Zhou, F., Rasmussen, A., Lee, S., Agaisse, H., 2013. The UPD3 cytokine couples environmental challenge and intestinal stem cell division through modulation of JAK/STAT signaling in the stem cell microenvironment. Dev. Biol. 373, 383–393.

Further Reading

Amoyel, M., Bach, E.A., 2012. Functions of the Drosophila JAK-STAT pathway: lessons from stem cells. JAK-STAT 1, 176–183.

CHAPTER

4

Induced Pluripotent Stem Cells

OUTLINE

LEARNING OBJECTIVES

1. Understand how cloning experiments led to our understanding of cellular reprogramming.
2. Explain gene regulation by transcription factors and other epigenetic factors.
3. Outline the role of transcription factors in reprogramming cells to a pluripotent state.
4. Discuss properties of the transcription factors used to induce pluripotency.
5. Describe the properties and use of viral vectors to reprogram cells.

https://doi.org/10.1016/B978-0-12-811014-0.00004-4

INTRODUCTION

Over the course of development, the cells of a multicellular organism become restricted to specific cell lineages, eventually adopting a final somatic fate. A cell's transition from pluripotency to a somatic state is determined by which genes it expresses. Pluripotent embryonic stem cells (ESCs) express genes that maintain self-renewal and pluripotency, while somatic cells do not. Instead, they express genes specific to the tissue or organ of which they are a part. In this chapter we will learn about genetic reprogramming and how an understanding of transcription factors and *epigenetics* enabled scientists to induce somatic cells to become pluripotent stem cells.

SOMATIC CELL NUCLEAR TRANSFER

In 1892, August Weismann proposed the Germ Theory (Weismann, 1892), the concept that genetic content is passed from one generation to the next. His theory was supported by Hans Spemann's 1938 experiments in which he separated an 8-cell-stage salamander embryo into individual cells that each cell grew into an adult. The Spemann experiment suggested that each cell had the same genetic material as the original embryo (Fig. 4.1) (Spemann, 1962). In 1952 Briggs and King began experiments where they transferred the nucleus of a blastula cell into an enucleated egg, both isolated from the frog, *Rana pipiens* (Briggs and King, 1952). Of the transfers, 40% resulted in normal tadpoles. The nuclear transfer experiment demonstrated that the nucleus of a blastula cell contains genetic material that is sufficient and competent to give rise to a normal tadpole. However, using the same technique they found that the nucleus of a cell from an embryo in the gastrulation phase of

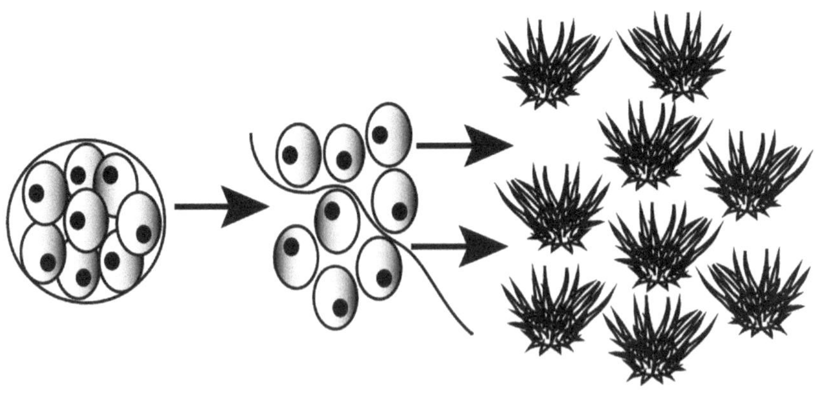

FIG. 4.1 Hans Spemann separated an 8-cell-stage sea urchin embryo with a strand of hair. Each cell went on to produce an adult sea urchin.

development had a much lower success rate (less than 20% of the transfers) of becoming a normal tadpole (Briggs and King, 1952). These experiments supported Weismann's Germ Theory, but the structure of DNA was still a year from being published, and the genetic differences between blastula versus gastrulation phase cells were not understood.

One of the questions driving these experiments was whether an irreversible change to the genetic material occurred during development that limited the cells' potential. Based on the work by Briggs and King, Sir John Gurdon did a series of nuclear transplant experiments using nuclei obtained from the aquatic frog *Xenopus laevis* at later and later developmental stages (Gurdon and Byrne, 2003). There are several advantages to using *Xenopus* over *Rana*. One is that egg laying can be induced with a hormone injection (Gurdon, 2013). Another is that its aquatic environment is easier to maintain in a lab than the terrarium required for *Rana*. Finally, *Xenopus* reach sexual maturity in less than a year, while *Rana* takes more than 3 years (Gurdon, 2013).

Gurdon first repeated the experiments of Briggs and King showing that a blastula stage nucleus transferred to an enucleated egg gave rise to a normal tadpole (Fig. 4.2). Gurdon repeated the experiment using nuclei from a gastrula stage nucleus, obtaining almost as many normal tadpoles as he had using a blastula stage nucleus. Gurdon repeated the experiment, each time using the nucleus from a cell obtained later in development. Eventually, he was able to use nuclei obtained from intestinal epithelial cells of feeding larvae transferred into an enucleated egg that could give rise to normal tadpoles (Gurdon, 1962). While the success rate was low, it demonstrated that cells isolated from somatic tissue could be reprogrammed to a pluripotent state. The results confirmed that there is no loss or irreversible change to DNA in somatic cells. The potential of a cell is therefore inherent in its genes regardless of its final fate. These somatic cell nuclear transfer (SCNT) experiments introduced the idea of *cellular reprogramming*.

EPIGENETICS

The next challenge was to define the differences between the gene expression of somatic and pluripotent cells. Every cell in a single organism has an identical copy of DNA. However, the genes *expressed* in nerve cells are different than the genes expressed in muscle cells or lung cells. Epigenetics refers to how gene expression is controlled in different cells by factors other than the gene sequence. During development, epigenetic factors dictate which genes are expressed, which in turn determines the fate of the cell. For example, different genes are expressed in a blastula stage cell than in a gastrula stage cell.

One of the most cited metaphors used to describe epigenetic regulation was first stated by Conrad Waddington. He compared an undifferentiated

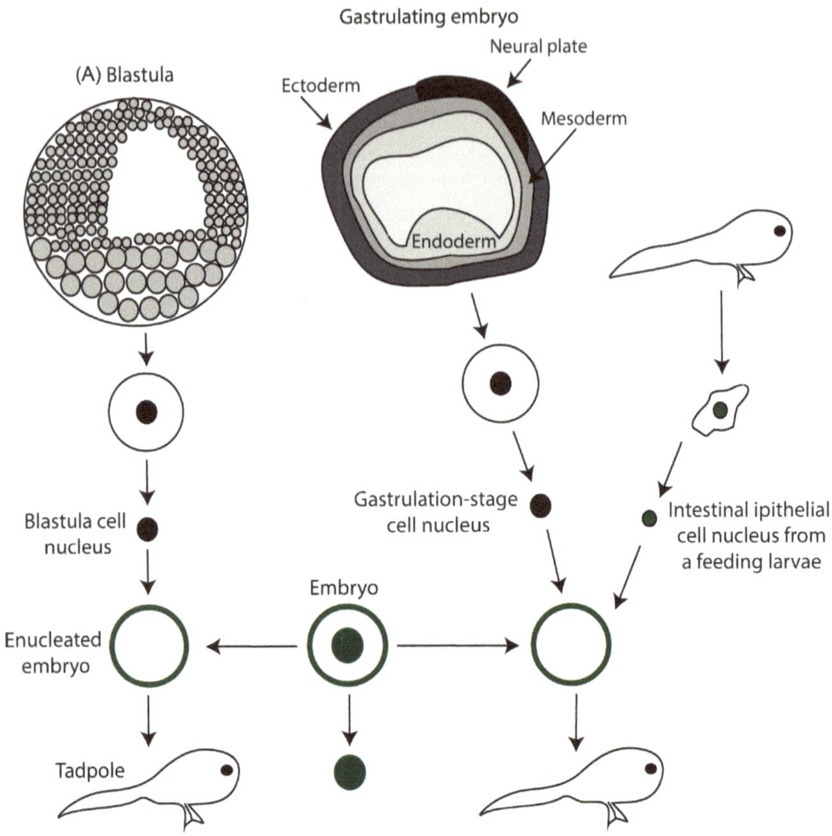

FIG. 4.2 The first somatic cell nuclear transfer (SCNT) experiments. Briggs and King, and later Gurdon, experimented with isolating nuclei from blastula stage cells and transferring them to enucleated embryos. Gurdon went on to transfer nuclei from cells in later and later stages of development. Using a nucleus from the intestinal epithelial cell of a feeding larvae, Gurdon was able to use SCNT to successfully obtain tadpoles.

cell to a ball at the top of a hill rolling through a sequence of bifurcating valleys separated by ridges (Waddington, 1957). Each new bifurcation represents an epigenetic mechanism with each choice or outcome contributing to the final fate of a cell. Once a selection is made, the ridges maintain the chosen fate. Importantly, our understanding of epigenetics has allowed researchers to roll the ball backwards and change the choice at specific bifurcations to bring about a new fate.

Chromatin

Within the cell, DNA is highly organized by protein structures called histones, which condense ~6 ft of genetic material into a single cell's nucleus (Ho and Crabtree, 2010). As we learned in Chapter 2, transcription factors are proteins that interact with DNA to produce RNA, which is translated into proteins (i.e., gene expression). The *chromatin* structure limits transcription factor access to DNA. These regions of the genome that are highly condensed and inaccessible for transcription are called heterochromatin. Regulatory elements are required that temporarily unwind DNA from its histones to create "open" chromatin called euchromatin that allows RNA polymerase to carry out gene transcription.

DNA is looped around a complex of eight histone proteins two times (147 base pairs), forming a *nucleosome*, with a linker (spacer) of DNA between each nucleosome and an additional histone protein called H1 (Fig. 4.3) (Tessarz and Kouzarides, 2014). Each nucleosome core consists of two copies of four histone protein types, H2A, H2B, H3, and H4 (Tessarz and Kouzarides, 2014). Six nucleosomes form a *solenoid*, a hexagon-like structure that organizes DNA into a compact unit. Solenoids are further twisted into loops, forming a structure that looks like a telephone cord, and this final structure is condensed into chromosomes. DNA that is coiled around the nucleosome core is inaccessible to transcriptional machinery. As such, regulating gene expression involves adjusting the interaction of DNA with histones. The coiling and compacting of DNA into chromatin requires epigenetic regulatory mechanisms dictated by a number of factors that can allow or inhibit the access of RNA polymerase to specific genes for transcription.

Histone Modification

Regions of euchromatin are regulated by posttranslational modification of histone proteins, an epigenetic mechanism that chemically modifies specific amino acids found in the tails of histone proteins (Fig. 4.4). The most common modifications are acetylation, methylation, or phosphorylation. Generally, acetylation enhances the unwrapping of DNA to an "open" euchromatin configuration allowing transcription to proceed. Methylation generally results in transcriptional suppression or gene silencing resulting in a "closed" configuration or heterochromatin. For example, in cancer cells tumor suppressor genes are hypermethylated resulting in unchecked cell growth (Baylin and Jones, 2011; Timp and Feinberg, 2013). Though the effect of phosphorylation can vary, phosphorylation of the H3 histone leads to nucleosome unwrapping and improves accessibility for transcription (Brehove et al., 2015). A length of chromatin may contain all of these posttranslational modifications that together result in a code interpreted by transcription factors or other DNA-associated proteins that regulate transcription (Jenuwein and Allis, 2001). Thus somatic cells have a different set of genes available for transcription compared to cells in the blastula stage or gastrulation stage of development, though the genomic DNA sequence remains unchanged.

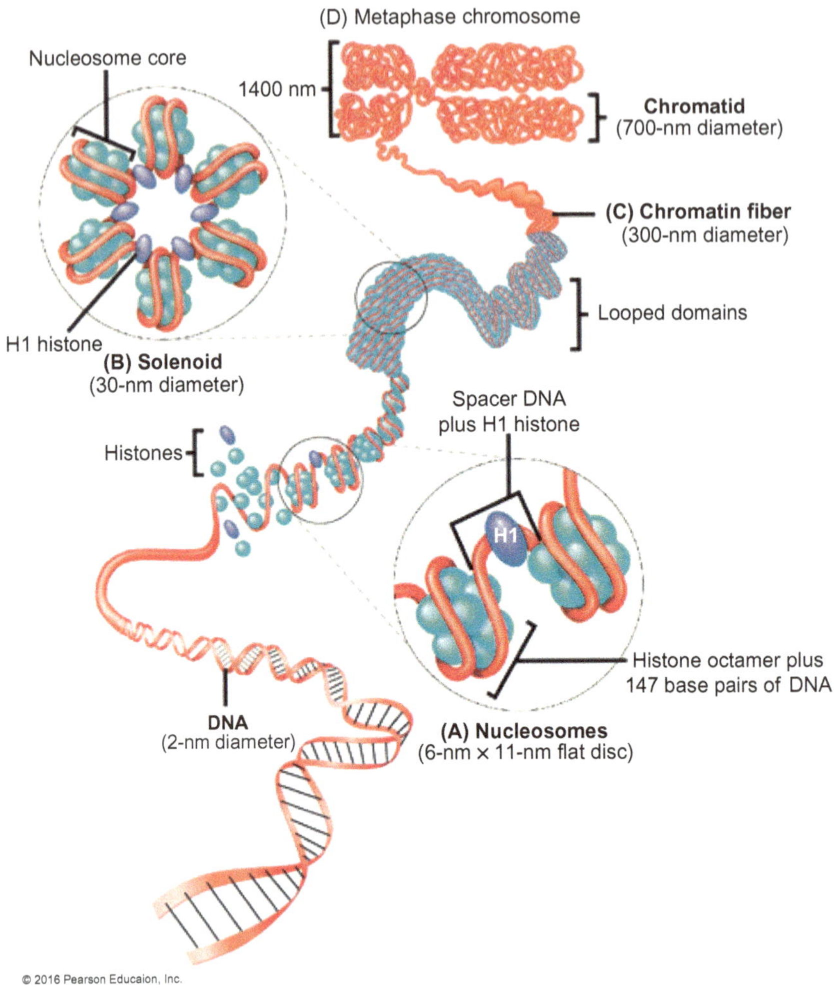

FIG. 4.3 The compact organization of DNA. Double-stranded DNA is wrapped around eight histone proteins with H1 histone separating each nucleosome. Nucleosome are organized into a hexagon-like solenoid, which are then looped into the compact chromatin structure. *Reproduced with permission from Klug et al., Essentials of Genetics, 9th ed., Pearson, 2017.*

Histone Remodeling

The position of DNA and histones are altered or remodeled through ATP-dependent mechanisms such as sliding, unwrapping, histone eviction or an exchange of histone variants (Fig. 4.5; Xu et al., 2013). Some transcription factors play a role in these processes by recruiting histone-modifying enzymes (Ho and Crabtree, 2010). As a cell undergoes differentiation the regulatory elements that make DNA available for

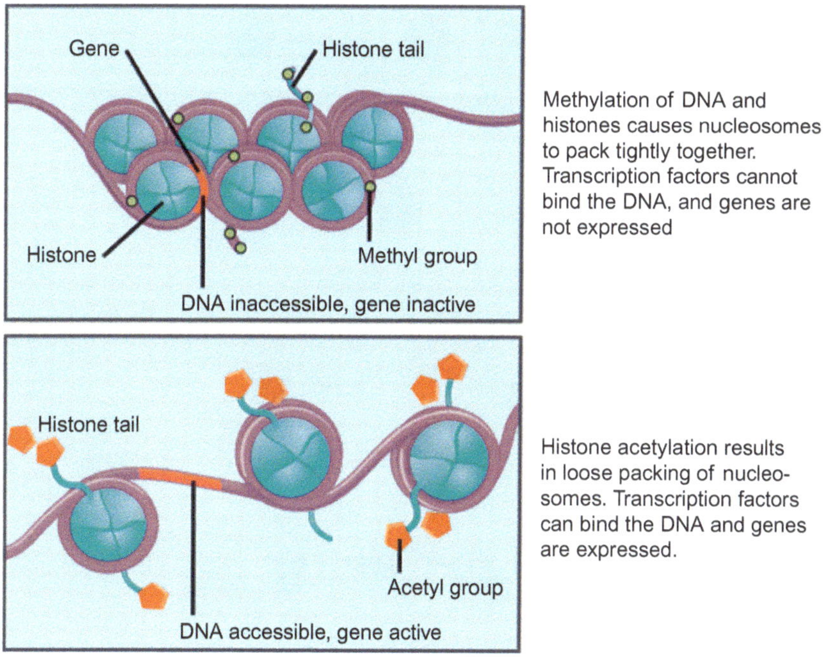

FIG. 4.4 Histone modification contributes to the regulation of DNA transcription. Methylation promotes the formation of heterochromatin (top). Genes present in heterochromatin are not accessible for transcription. Acetylation promotes the formation of euchromatin (bottom) that allows the transcription of genes in these regions. Credit: Biology, 16.3-2 Eukaryotic Epigenetic Gene Regulation by CNX OpenStax, licensed under Creative Commons Attribution 4.0 International.

transcription factor binding are a dynamic part of epigenetic regulation (Voss and Hager, 2013). Thus genes that contribute to cell fate are part of the euchromatin. For example, if a cell starts down the path to an ectodermal fate, chromatin remodeling provides access to ectodermal genes while preventing access to endodermal or mesodermal genes.

Transcription Factors

Transcription factors recognize and bind to specific DNA sequences called promotor or enhancer sequences. Promotor sequences are typically upstream of a gene and ensure accurate initiation of transcription. Some transcription factors are generalized to many promoters, while other transcription factors are site-specific to one promotor and gene. When a transcription factor binds to the promotor sequence, it recruits general transcription factors and mediator proteins that in turn recruit RNA polymerase; together these proteins form the transcription initiation complex (Fig. 4.6). Other transcription factors called

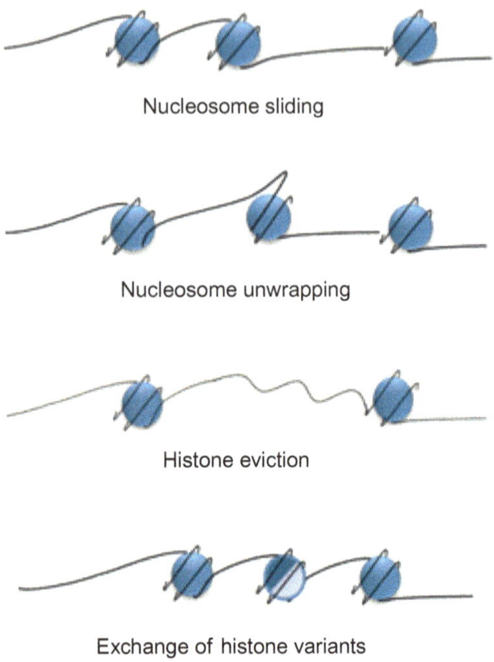

FIG. 4.5 Histone remodeling includes the mechanisms of nucleosome sliding and unwrapping, histone eviction or the exchange of histone protein variants. These epigenetic processes allow other proteins such as transcription factors to interact with the DNA. *Credit: Xu, Y.Z., Kanagaratham, C., Radzioch, D., 2013. Chromatin remodelling during host-bacterial pathogen interaction. In: D. Radzioch (Ed.), Chromatin Remodelling. InTech. Licensed under CC BY 3.0.*

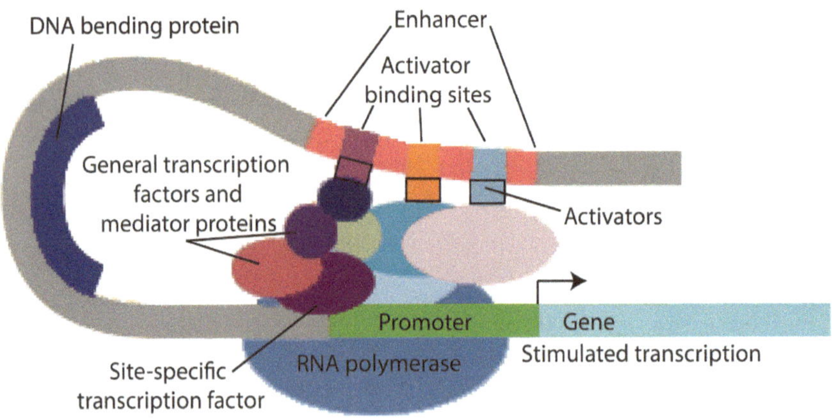

FIG. 4.6 Transcription factors interact with sequences of DNA to promote and enhance transcription. Activating transcription factors are epigenetic regulators of specific genes. DNA bending protein brings the activators into the proximity of the promotor where additional transcription factors are bound and interact with the activators. Transcription factors at the promotor recruit RNA polymerase and transcription can proceed.

activators target enhancer sequences that initiate the bending of the DNA to allow interaction with the proteins bound to the promoter region. A typical enhancer sequence can be up to 500 base pairs in length and contain multiple binding sites for at least two or three different activators (Levine and Tjian, 2003). An enhancer sequence is found hundreds to thousands of base pairs away, either upstream or downstream from the gene it regulates. DNA bending in tightly packaged chromatin brings the enhancers within close approximation of the promoter and gene, often over great distances (500 base pairs to thousands of base pairs away), further ensuring transcription.

Our ability to reprogram cells is based on an understanding of how epigenetic factors combine to control cell fate in the body. Early studies in *Drosophila* showed that a single transcription factor gene expressed during development could determine the fate of cells to become either legs or antennae (Schneuwly et al., 1987). In a famous experiment, scientists altered transcription factor expression levels to show how loss-of-function could convert the second leg pair into antenna, while gain-of-function led to the conversion of antenna to legs (Fig. 4.7). Another group demonstrated how overexpressing a single gene into fibroblasts could convert them to myoblasts in culture (Davis et al., 1987). These experiments demonstrated that transcription factors can determine cell fate by expression within spatial regions of the embryo, and temporal stages of expression during development.

In summary, epigenetic mechanisms underlie the spatial and temporal expression of genes by regulating histone remodeling and modification, the chromatin structure ("open" vs. "closed" DNA), and the interactions of transcription factors between enhancers and/or promoters with target

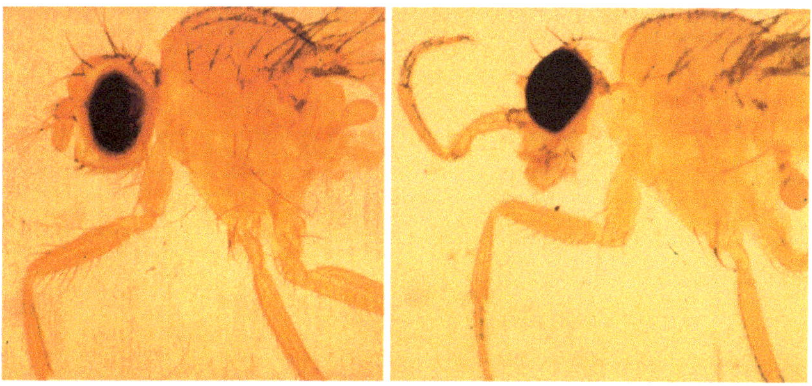

FIG. 4.7 Head and thorax of a wildtype fly (left) compared to a mutant fly in which the altered expression of a single gene *Antennapedia* at a defined larval stage converts the antenna to a leg (right). *Figure used with permission from Morata, G., Sánchez-Herrero, E., 1998. Pulling the fly's leg: developmental biology. Nature 392, 657–658.*

genes over the course of development (Voss and Hager, 2013). These advances led to a better understanding of how somatic cells can be reprogrammed to a pluripotent state.

INDUCED PLURIPOTENT STEM CELLS (iPSCs)

As an understanding of epigenetic mechanisms increased, scientists began investigating how to control the expression of specific genes. In the early 2000s, scientists were focused on which transcription factors were active in proliferating ESCs. By 2006 more than a dozen studies had identified genes and transcription factors that were expressed in self-renewing ESCs or tumors (Takahashi and Yamanaka, 2006). Several transcription factors were shown to be important for maintaining pluripotency in ESCs and early embryos, including transcription factors upregulated in tumors (Boiani and Schöler, 2005).

Dedifferentiation

In 2006 the lab of Dr. Shinya Yamanaka experimented with a number of transcription factors known to be expressed by stem cells. They used a mouse line that was engineered so that cells in which the reactivated ESC-specific gene Fbx15 could be isolated by antibiotic resistance (Takahashi and Yamanaka, 2006). The literature suggested that 24 transcription factor genes were pivotal in the maintenance of ESC identity. Takahashi and Yamanaka used a *retrovirus* to transfect embryonic fibroblasts with all 24 factors, resulting in teratoma-like colonies that had reactivated the Fbx15 reporter (Box 4.1). The cells resembled ESCs and could propagate indefinitely. Next, they narrowed their transcription factor candidates by removing one factor at a time from the pool of 24. By this process, they identified four transcription factors, Oct3/4, Sox2, c-Myc, and Klf4, which as a group were necessary and sufficient to obtain ESC-like colonies (Box 4.2). In culture, clones from the colonies proliferated indefinitely and expressed genes associated with self-renewal and pluripotency. The teratomas contained an array of tissue types (cartilage, muscle, adipose, epithelium and CNS tissue) representing all three germ layers, further confirming pluripotency.

Again using the factors Oct3/4, Sox2, c-Myc and Klf4, the Yamanaka group next transfected fibroblasts taken from the tail tips of adult mice (7–12 weeks, including a GFP-expressing animal) and were able to isolate more than a dozen antibiotic resistant colonies from which they obtained six clones (Takahashi and Yamanaka, 2006). In culture, the clones from transfected adult cells proliferated similarly to the transfected embryonic fibroblasts and had a similar gene profile. Next, they injected the

BOX 4.1

SPOTLIGHT ON VIRUSES

Retrovirus, Lentivirus, and Adenovirus

The potency of a virus to infect cells, or tropism, depends on viral envelope or capsid protein interactions with viral receptors on the host (target) cells (Fig. 4.8). Viral envelopes and capsids can be ecotropic (narrow range of targets), amphotropic (can target most mammalian cells) or polytropic (indiscriminately infectious). Much research has gone into engineering viruses in order to increase cell-targeting specificity to introduce DNA into the cells. One method to accomplish this is through pseudotyping, in which the capsid of nonenveloped vectors or surface glycoproteins of enveloped viruses is altered. Another area of investigation is the use of adaptor molecules that recognize the virus on one side and the target cells on the other. However, adaptors have not been explored for clinical use. Another approach is to insert a ligand in the virus surface that triggers endocytosis or membrane fusion when it is bound by the target cell receptor.

Once inside a cell, steps to replicate the virus and produce viral proteins begin. Viruses have a simple genome that carries the genes required for replication (described later). Viruses mainly use the host cell's transcription or translation machinery, though some have proteins within the capsid that assist with initial replication. Some viruses integrate their genes into the host cell's DNA and only proliferate in

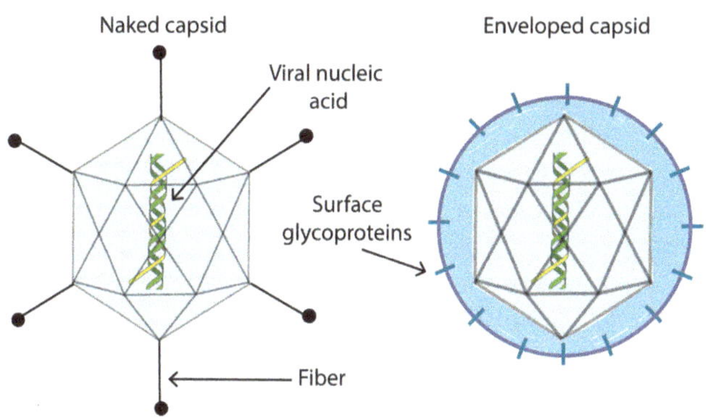

FIG. 4.8 Viral vectors are used to induce somatic cells by introducing genes that, when expressed, will induce pluripotency. Left, a naked capsid is shown with adenoviral fibers. The naked capsid is ~20 nm in diameter. Right, the enveloped capsid representing retroviruses and lentiviruses. Including the protruding surface glycoproteins the diameter is ~300 nm.

Continued

BOX 4.1

SPOTLIGHT ON VIRUSES—CONT'D

dividing cell populations. Other viruses remain *extrachromosomal*, which makes them infectious to dividing and somatic (nonproliferating) cells. The three virus types described in the following paragraphs have unique properties that enhance or detract from their efficacy as a viral vector for inducing pluripotent stem cells, reprogramming somatic cells, or for gene therapy.

Retroviruses have a simple RNA genome that consists of two identical RNA molecules connected by a dimer linkage structure. The retrovirus capsid is enclosed in a lipid envelope expressing proteins that bind to receptors on a potential host cell, typically a macromolecule, which initiates entry of the virus. The envelope fuses with the cell membrane allowing the capsid to enter the cell. The retrovirus must be uncoated from the capsid, while concurrent reverse transcription is carried out by a protein packaged within the capsid called reverse transcriptase, making complementary DNA from the viral RNA (Hu, 2014; Nisole and Saib, 2004). The resulting DNA is transported into the cell's nucleus and integrated into the chromosomal DNA, making it heritable in dividing cells.

The retrovirus genome is simple, containing three genes from which several proteins are encoded by alternative splicing. The *pol* gene encodes reverse transcriptase, integrase and protease enzymes (Voevodin and Marx, 2009). The *gag* gene encodes structural proteins and the *env* gene encodes envelope proteins (Kotterman et al., 2015). Retroviral vectors can be modified to carry a gene of interest that is transferred to a target cell, called a transgene. Once the transgene is integrated into the target cell DNA, the gene can be passed on to daughter cells. Typically, viruses used to transfect cells have the viral protein-coding genes deleted to inhibit further infection (Bouard et al., 2009).

An early observation was that retroviral integration activated proto-oncogenes (Bouard et al., 2009). The use of "insulator sequences" has helped prevent the transgene integration event from affecting the upregulation of surrounding genes as well as the silencing of the transgene, which can occur over time (Schambach et al., 2007). Finally, current research is focused on how to control the site of retroviral integration in order to limit the potential for insertional mutations or other complications due to integration (Bouard et al., 2009). Retroviruses have a preference for sites where active gene transcription occurs (Kotterman et al., 2015). Thus the clinical use for retroviruses to reprogram cells or introduce a corrected gene carries the risk of insertional mutagenesis, tumorigenesis and incomplete pluripotent gene silencing (Simara et al., 2013).

BOX 4.1

SPOTLIGHT ON VIRUSES—CONT'D

Lentiviruses are a subtype of retrovirus. Lentiviruses are capable of infecting nondividing and actively dividing cells, making it useful for targeting somatic cells. Similar to retroviruses, they have two identical RNA strands enclosed in an envelope. Its genome carries the *pol, gag and env* genes as well as six genes that encode accessory proteins (Kotterman et al., 2015). Improvements to the safety of lentivirus vectors have included removal of the accessory protein genes and the sequences flanking the *pol, gag and env* genes called long terminal repeats. Lentiviruses have a lower risk of insertional mutagenesis due to integration preferences (Montini et al., 2009). Lentiviruses are susceptible to attack by the human immune system and are more useful for ex vivo gene delivery (DePolo et al., 2000). Despite this drawback, lentiviruses have potential for safe gene therapy use (Kotterman et al., 2015).

Adenoviruses do not have an envelope; instead, only a capsid made up of proteins surrounds its double-stranded DNA genome. The capsid is an icosahedral (20 faces) with a fiber that projects from each vertex; the fiber serves as the ligand for cellular receptors (Zhang and Bergelson, 2005). Adenoviruses are divided into six subgroups (A–F) and some correlation between subgroups and specific organ systems has been observed: B1, C, and E subgroup infections affect the respiratory system; B2 causes kidney and urinary tract infections; F causes gastroenteritis; D results in epidemic keratoconjunctivitis (Zhang and Bergelson, 2005).

Infection with an adenovirus typically results in short-term expression because they are nonintegrative and induce an immune response that results in the destruction of cells that host the virus (Bouard et al., 2009). During translation alternative splicing is used in the sense and antisense orientation to encode proteins (Kotterman et al., 2015). For safety, the adenoviral genes are typically deleted (gutless) to inhibit further infection and lower immune responses. To ensure production of the virus a helper plasmid containing the genes necessary to encode the replication machinery is provided in *trans*, which means introduced to the cells with the adenovirus on a separate plasmid. Adenoviruses have a capacity for large transgene inserts and can also be easily purified at high concentrations (Lai et al., 2002). The nonintegrative, extrachromosomal nature of adenoviruses, ease of inserting transgenes and transfecting cells has made them a popular choice for creating iPSCs or transducing somatic cells (Hu, 2014). Some drawbacks are that they have a low efficiency for reprogramming cells, and the transgene expression is temporary (Hu, 2014). Another issue is the broad immune response induced by adenovirus particles

Continued

BOX 4.1

SPOTLIGHT ON VIRUSES—CONT'D

(adenovirus or empty adenovirus capsid) (Stilwell and Samulski, 2004). Another group reported a high level of background integration that is an issue for using adenoviruses for clinical applications (Harui et al., 1999). However, since the unfortunate death of Gelsinger in 1999, much work has gone into improving the safety of adenoviruses for clinical trials, and as of 2011 up to 16,000 patients have received an adenovirus in a clinical setting (Pesonen et al., 2011).

Adeno-associated viruses (AAV) are increasingly used as vectors in gene therapy clinical trials. The change between the increased use of AAVs from 2009 to 2017 (3.7%) is similar to the decrease in the use of adenoviruses (4.3%; Edelstein, 2017). AAVs have an ability similar to adenoviruses to infect both dividing and postmitotic cells (Lai et al., 2002). The benefits of AAVs is that they have low background integration and have the potential to be a good in vivo gene delivery vector. AAVs have a single-stranded DNA genome enclosed in an icosahedral capsid. Without the presence of an adenovirus AAVs are unable to replicate (Daya and Berns, 2008). Gutless AAVs with the replication machinery provided in *trans* can transduce dividing and postmitotic cells. Once an AAV has entered a cell and a second strand of DNA is produced, it forms a circular DNA structure that can remain in postmitotic cells for years (Kotterman et al., 2015).

BOX 4.2

SPOTLIGHT ON CELLULAR REPROGRAMMING FACTORS

Here we provide a brief description of six important factors used to induce somatic cells to a pluripotent state.

Sox2 stands for SRY (sex-determining region Y)-box 2 and is a member of the Sox gene family, of which more than 20 genes have been identified in mouse and human (Schepers et al., 2002). Sox2 is essential for normal embryonic development. It has diverse functions due to the combination of various cofactors that influence the binding of Sox2 to its target genes (Zhang, 2014). Many Sox2 target genes, including those that affect Oct4regulate pluripotency, making it a critical requirement for inducing pluripotency in somatic cells (Kashyap et al., 2009; Masui et al., 2007, 2007).

BOX 4.2

SPOTLIGHT ON CELLULAR REPROGRAMMING FACTORS—CONT'D

Oct4 stands for octamer-binding transcription factor 4. Also known as Oct3/4, it is an important transcription factor involved in critical early cell fate decisions and a regulator of genes that coordinate multiple cellular functions (Loh et al., 2006). For example, removing Oct4 from pluripotent cells caused them to revert to trophoblasts, the cells that surround the developing blastocyst. Conversely, overexpression of Oct4 in ESC results in endodermal differentiation (Niwa et al., 2000). Oct4 often binds in partnership with Sox2 and many pluripotency genes including *Nanog* are regulated by an enhancer that contains a binding site for Oct4 and Sox2 (Loh et al., 2006; Masui et al., 2007). Further, in the presence of valproic acid, a histone deacetylase inhibitor, induced pluripotency was achieved with Oct4 and Sox2 alone (Huangfu et al., 2008).

Klf4 belongs to the family of Krüppel-like factor (Klf), zinc-finger proteins (Nakagawa et al., 2008). Klf4 is both a transcriptional activator and repressor of genes that regulate the cell cycle (Rowland et al., 2005). During stress or normal differentiation, Klf4 is expressed at high levels and proliferation is inhibited (Black et al., 2001). The loss of Klf4 can lead to cancer, in good agreement with its role as an inhibitor of proliferation (Tetreault et al., 2013). However, high levels of Klf4 have also been associated with cancer (Tetreault et al., 2013). A study by Rowland et al. (2005) found that the expression of other genes determine whether Klf4 will stimulate or inhibit cell proliferation. Klf4 is important for initiating reprogramming and maintains pluripotency in iPSCs (Hawkins, 2014; Takahashi and Yamanaka, 2006). Klf4 targets genes in tandem with Oct4 and Sox2 that have the highest impact on regulating pluripotency (Fig. 4.8) (Sridharan et al., 2009).

c-Myc is a transcription factor that regulates gene expression to establish pluripotency while balancing energy expenditures with metabolism (Dang, 2012; Varlakhanova et al., 2010). As a proto-oncogene c-Myc was later shown to be dispensable to induce pluripotency, and its absence reduces the tumorigenicity of iPSCs (Okita et al., 2007; Wernig et al., 2008). While cell proliferation is slower without c-Myc, the rate of tumor formation in *chimeric* mice derived from iPSCs is significantly reduced (Nakagawa et al., 2008). In mouse fibroblasts c-Myc is important for early downregulation of fibroblast-specific genes (Sridharan et al., 2009). However, in hESCs, induction of c-Myc resulted in apoptosis or differentiation (Sumi et al., 2007).

Continued

BOX 4.2

SPOTLIGHT ON CELLULAR REPROGRAMMING FACTORS—CONT'D

Nanog is a transcription factor whose actions are critical during development in order for the inner cell mass to mature into pluripotent epiblast cells (Mitsui et al., 2003). Factors such as Sox2 and Oct4 are necessary for the initial steps to induce pluripotency such as loss of differentiated characteristics. Nanog is then important during later stages for acquiring and stabilizing a pluripotent state (Silva et al., 2009; Sun et al., 2014).

Lin28 is an RNA binding protein that inhibits molecules that lead to maturation (Jin et al., 2011; Rybak et al., 2008). Lin28 is important for self-renewal, but when overexpressed can be oncogenic (Viswanathan et al., 2009). Two isoforms, LIN28A and LIN28B, have separate roles during reprogramming. LIN28B is involved in early reprogramming, while LIN28A is involved in later stages of the cells becoming pluripotent (Zhang et al., 2016).

induced pluripotent stem cells (iPSCs) from one of the GFP-positive clones into wild-type mouse blastocysts to determine whether the iPSCs could be incorporated into developing embryos. Indeed, they obtained 18 pups that had GFP-positive cells in all three germ layers confirming the pluripotency of their iPSC clones.

The process of inducing somatic cells to become pluripotent stem cells is called *dedifferentiation* as it causes somatic cells to revert to an earlier developmental stage. As discussed previously, dedifferentiation is accomplished by either SCNT or by controlling transcription factor expression. Pluripotent stem cells have been derived from a multitude of cell types using the transcription factors published by the Yamanaka lab to induce pluripotency. Neural cells, Schwann cells, astrocytes, keratinocytes, melanocytes and adipose cells have all been induced to derive pluripotent stem cells, though fibroblasts still remain a common cell type to derive iPSCs (Aasen et al., 2008; Eminli et al., 2008; Masaki et al., 2013; Ruiz et al., 2010; Takahashi and Yamanaka, 2006; Utikal et al., 2009; Zapata-Linares et al., 2016).

The fate of cells previously assumed to be fixed can now be genetically reprogrammed. Further, in a culture dish one cell type can be reprogrammed into another type of cell by a process called *transdifferentiation*, specifically when the conversion is to a cell type of another lineage (e.g., endodermal to mesodermal) (Hochedlinger and Jaenisch, 2015). Transdifferentiation was first demonstrated by Davis et al. (1987) when they used a retroviral *vector* to express the muscle-specific transcription factor MyoD in fibroblasts. The transfected fibroblasts were stably converted to myoblasts, and in some cases, the authors observed myogenesis. Of interest, MyoD shares a region similar to c-Myc, one of the transforming factors used by the Yamanaka lab to obtain iPSCs from mouse fibroblasts.

Advances in transdifferentiation have given rise to the concepts of direct versus indirect programming. Direct reprogramming is the process of transdifferentiating a somatic (differentiated) cell into another type of somatic cell. Indirect reprogramming is when a somatic cell is first dedifferentiated to a pluripotent state, and then reprogrammed to the desired somatic cell type (Nicholas and Kriegstein, 2010). Studies are now focused on the clinical applications of cellular reprogramming. For example, mammalian retinal pigment epithelium can be reprogrammed to give rise to photoreceptor-like cells by the transfection of photoreceptor-specific transgenes (Yan et al., 2013). Another possible experiment would be to introduce genes into retinal pigment epithelium cells in the eye to reprogram them directly into photoreceptor cells. In another study, lymphoblastic leukemia cells were reprogrammed into nonleukemic macrophages, laying the groundwork for developing a new treatment for leukemia (McClellan et al., 2015). An advance in cellular reprogramming by Pagliuca et al. (2014) identified the extracellular factors necessary to reprogram human ESCs (hESCs) in vitro through the modulation of multiple signal transduction pathways without the use of transgenes. The study generated functional human pancreatic ßcells, the cells that produce insulin for patients suffering from diabetes (Mfopou and Bouwens, 2015; Pagliuca et al., 2014). The research studies described here could result in major benefits to patients affected by these diseases.

It was observed by many that if human somatic cells could be reprogrammed to a pluripotent state, one would have an unlimited source of patient-specific cells. Human iPSCs could have the potential and promise of hESCs without any of the associated ethical concerns. Likewise, concerns regarding the enucleated egg used for SCNT would no longer be necessary. Research on how the reprogramming factors affected human cells began immediately.

One of the initial studies introduced the c-Myc gene in cultured hESCs (Sumi et al., 2007). The hESCs confirmed to express c-Myc began to apoptose or undergo differentiation. A study from the Yamanaka lab found that 20% of mouse fibroblast clones developed tumors when c-Myc was expressed and that c-Myc retroviral expression occasionally reactivated, causing the tumors (Okita et al., 2007). The authors concluded that for clinical uses c-Myc should be avoided when deriving iPSCs (Okita et al., 2007). Of note, another finding was that using a different transcription factor called Nanog to select for inducible cells (instead of Fbx15), produced iPSCs that were more similar to ESCs (Okita et al., 2007).

Finally, the end of 2007 saw the induction of human fibroblasts into pluripotent stem cells by the Yamanaka lab using the same four pivotal genes: Oct3/4, Sox2, Klf4, and c-Myc with a retroviral system (Takahashi et al., 2007). A few weeks later James Thomson, the same author of the paper describing the first derived ESCs from human blastocysts (Chapter 1) also induced pluripotent stem cells from human somatic cells (Yu et al., 2007). The Thomson lab used a lentiviral system to *transduce* cells to express Oct4 (another name for Oct3/4), Sox2, Nanog, and Lin28. Both studies showed that the human iPSCs expressed markers of ESCs and could differentiate into cell types from all three lineages.

The key characteristics of iPSCs are that they express the same genes and proteins as stem cells, can form tumors in vitro or in vivo that contain cells from all three germ layers, and can contribute to many different tissues when injected into mouse embryos at a very early stage in development (NIH Stem Cell Information, 2016). The key characteristics of stem cells are that they can proliferate indefinitely, self-renew, express pluripotency markers, and differentiate into cells from all three germ layers (Daley et al., 2009). Under these definitions it is agreed that hESCs, human iPSCs, and human AFSCs (Chapter 2) all qualify as pluripotent (Daley et al., 2009). However, studies found differences in gene expression depending on the methods used to culture and passage ESCs and iPSCs (Hohenstein Elliott et al., 2012). Different reprogramming methods give rise to a wide variety of iPSCs, each of which must be carefully characterized before they can be considered for clinical applications (Alvarez et al., 2012). Indeed, studies have found genetic variations in some iPSCs that could be oncogenic in nature (Blasco et al., 2011).

ESCs were touted as an unlimited supply of cells that could be differentiated into any somatic cell of the body. For debilitating injuries and disease- or age-related degenerative disorders, ESCs could serve as healthy replacement cells. Initial studies focused on optimizing conditions to differentiate ESCs into a desired cell type (Pedersen, 1994). A major hurdle to using ESCs therapeutically is that as ESCs differentiate they begin to express immune system-sensitive markers called histocompatibility complex (MHC) molecules that signal the presence of foreign cells in the host resulting in immune rejection (Fairchild et al., 2005). Although ESCs do not initially express MHC molecules, it is not recommended to directly transplant ESCs into a patient because of the tendency of ESCs to form tumors, which is another hurdle to using ESCs therapeutically (Fairchild et al., 2005). Further, the use of ESCs is embroiled in ethical arguments that have reached into the political arena. ESC research is currently legal, but when using federal research dollars, only on certain existing cell lines (Murugan, 2009).

The ability to establish pluripotent stem cells from somatic cells opened two major avenues of research for human disease (Fig. 4.9).

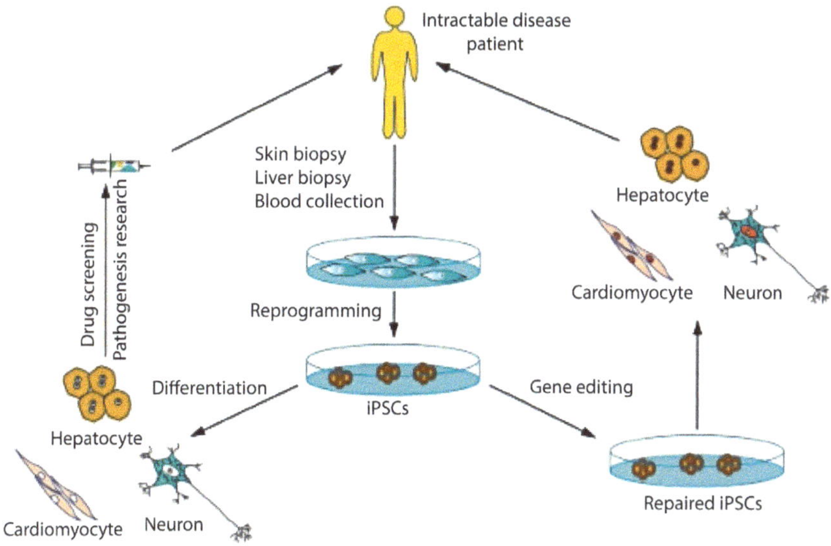

FIG. 4.9 The discovery of iPSCs has allowed advances in stem cell technologies to proceed in two main directions. Somatic cells from a patient are induced (center) into cultured pluripotent cells that are then used (left) to differentiate the cells into a desired cell type to understand the pathophysiology of the disease, and test potential drugs or other therapies to treat the disease. The second area of research (right) works to correct the disease-causing mutation in the cultured iPSCs from the patient using gene therapy, differentiate the treated cells into the needed cell type(s) and inject them back into the patient. These areas of investigation avoid the use or destruction of an embryo. *Modified from Chun, Y.S., Byun, K., Lee, B., 2011. Induced pluripotent stem cells and personalized medicine: current progress and future perspectives. Anat. Cell Biol. 44, 245; under the terms of the Creative Commons Attribution Non-Commercial License (http://creativecommons.org/licenses/by-nc/3.0/).*

The first is that somatic cells from a patient can be induced into cultured pluripotent cells that are then used to understand the pathophysiology of the disease, and to test potential drugs or other therapies to treat the disease. The second area of research works to correct the disease-causing mutation in the cultured iPSCs from the patient using gene therapy, a treatment method that attempts to replace mutated or missing genes that caused the disease or disorder, with a normal (corrected) gene. Cells are treated in vitro, differentiated into the needed cell type(s) and injected back into the patient. These areas of investigation avoid the use or destruction of an embryo, unlike research on ESCs or pluripotent cells produced by SCNT. The discovery of iPSCs has allowed advances in stem cell technologies to proceed much faster without the weight of ethical issues surrounding the use of human fetal tissue or eggs. However, the risks associated with viral vectors used to induce pluripotency and the required careful characterization of reprogrammed iPSCs has delayed the progress of iPSC use in the clinic.

The same viral vectors used to transduce reprogramming factors into cells can also be used for gene therapy. The first gene therapy trial in 1995 successfully used a retrovirus for two patients with severe combined immunodeficiency (SCID; also known as "bubble boy" disease) due to a genetic defect in a T cell enzyme that prevents normal immune system function (Blaese et al., 1995). T cells from the two patients with the mutation were cultured and transduced with a retrovirus containing the corrected gene. The transduced T cells were then reinfused back into the patients. After 2 years the gene therapy treatment was halted when normal enzyme activity in the patient's blood was observed. Two years after that, the corrected gene was still present in the patient's T cells and both patients were able to attend public school, now that their immune system function was restored (Blaese et al., 1995). The conclusion was that gene therapy using a retroviral vector was safe.

Four years later, another gene therapy case tested the safety of a treatment for infants born with a fatal form of ornithine transcarbamylase (OTC) deficiency. Jesse Gelsinger was an 18-year-old patient with a nonlethal form of the disease who volunteered to participate in the study. Gelsinger and 17 other participants received an infusion of corrective genes encased in a disabled adenovirus (Raper et al., 2002). Previous tests with the same dose on mice, nonhuman primates and one human patient indicated that flu-like side effects and mild liver inflammation may occur, but would subside without treatment. Within a day of the infusion, Gelsinger presented with jaundice and altered mental status. Three days later on September 17, 1999, Gelsinger died from multiple organ failure (Raper et al., 2003). An autopsy found that the vector had activated his immune system and led to the acute release of inflammation mediators, most likely in response to proteins on the vector capsid (Wilson, 2009). At that time, not only did plans for gene therapy come to an end, but at some institutions all clinical therapy ceased until greater oversight on human research subjects could be established (Wilson, 2009).

In 2000, the Office for Human Research Protections (OHRP) was established. While the Gelsinger case was important, there had been other unregulated research on humans, including the use of radiation, which led to the timely launching of the OHRP. One of the important roles of the OHRP is oversight of institutional review boards (IRBs), whose members review applications to conduct research on human participants. Any institution that has research being conducted on humans must have an IRB. An IRB consists of at least five members who are qualified to evaluate research but are not involved in the study. Ethical guidelines of the OHRP require respect for persons, including informed consent, beneficence and justice. Using the guidelines of the OHRP, clinical research can proceed only after it has been determined that the study is both safe for and a benefit to the patient.

The emergence of iPSC technology makes it possible to safely test viral vectors on patient-derived, cultured cells to test gene therapy treatments.

As such, new treatments for disease are being made possible through advances that safely introduce the normal or corrected gene to cells. In the last 10 years, more than 1200 gene therapy clinical trials have been approved worldwide (http://www.abedia.com/wiley/index.html). The largest number have targeted cancer, with the second largest targeting monogenic diseases, those caused by a mutation to a single gene, followed by infectious, cardiovascular, neurological and ocular diseases. However, until 2017 none of the gene therapy clinical trials had successfully used iPSC reprogrammed from patients as part of the actual treatment.

In a groundbreaking study, a group of researchers in Masayo Takahashi's lab in Japan made human iPSCs from a patient suffering from age-related macular degeneration. The dermal fibroblast cells were reprogrammed to differentiate into retinal pigment epithelial (RPE) cells using a viral vector (Jin et al., 2012; Kamao et al., 2014; Mandai et al., 2017). The RPE cells were manipulated in culture until they formed a thin sheet that could be transplanted back into the eye of the donor (Mandai et al., 2017). Before the surgery to transplant the RPE sheet, cells were screened exhaustively for proper integration of the vector, mutations or alterations to the genome, and tumorigenicity. One year after the surgery, the patient's vision remained stable, and by some measures was slightly improved. This was accomplished without the patient having to take immunosuppressants, as is the case when ESCs are used. Takahashi's research group has recently taken the next step toward making iPSC therapy more readily available. Using iPSCs derived from a donor, they used the same induction and screening methods to derive an RPE sheet that was then transplanted into an unrelated patient (Cyranoski, 2017). The recipient of the allogeneic RPE sheet is part of an ongoing clinical trial in which five other patients will receive the treatment using the donor iPSCs. This study is presently creating the standards by which iPSCs can become a valuable and more readily available gene therapy tool.

Gene editing has traditionally been accomplished using homologous recombination, a method requiring the creation of a protein called a zinc finger nuclease (ZFN) that could recognize up to 36 base pairs of a DNA sequence. The ZFNs are paired with a nonspecific *FokI endonuclease*, which creates double-stranded DNA breaks. The ZFNs bind to the DNA at its target sequence, and then FokI cleaves the phosphodiester bonds between the nucleotides, allowing a transgene to be inserted into the break site or removal of a specific sequence.

A new technology called CRISPR/Cas9 is making gene editing easier and more accurate. Instead of piecing together ZFNs to recognize a sequence, CRISPR/Cas9 only requires a guide RNA sequence of 17–20 nucleotides to direct the Cas9 enzyme where to cleave DNA. Presently, CRISPR/Cas9 has been modified to provide three basic functional options. The first modification allows CRISPR/Cas9 can make double-stranded breaks in the DNA for the insertion or deletion of a new sequence. Another version only makes single-stranded breaks after which precise replacement or repairs can be made. The third version has been called "broken scissors," which binds to the target sequence, but does not cut the DNA. Instead, this version blocks other proteins from interacting with DNA at that sight resulting in gene silencing. The broken scissors version has been further modified with activating proteins that stimulate expression of a specific gene or epigenetic modulators that carry out histone modification. CRISPR/Cas9 has provided a faster, more accurate method to edit genes with increased editing options that is already seeing clinical progress.

Using iPSCs, progress is also being made in other research areas such as Parkinson disease (PD), a neurodegenerative disorder that affects more than 10 million people a year worldwide (Parkinson's Foundation, http://parkinson.org/Understanding-Parkinsons/Causes-and-Statistics/Statistics). Recently, a study used human iPSCs to derive the dopaminergic neurons that degenerate in PD patients (Kikuchi et al., 2017). These neurons were grafted into the putamen of macaques that had been treated with MPTP, a synthetic heroin-like drug that selectively damages dopaminergic neurons. For the next 12–24 months the treated animals were monitored using brain imaging techniques. The grafted cells appeared to survive and function without tumor growth or an immune response, and significant improvement in neurological and spontaneous movement scores over control animals (vehicle transplant) were observed. After 2 years, the animals did not appear to have any adverse effects from the transplant. Histological analysis of the brain revealed no evidence of tumor formation. The neurons had normal morphology and dopaminergic marker expression. These data demonstrate the potential of iPSCs for the treatment of PD and other neurodegenerative disorders.

SUMMARY

The discovery of how to induce pluripotent cells from somatic cells revolutionized stem cell research. Shinya Yamanaka and Sir John Gurdon shared the Nobel Prize in 2012 for independent efforts demonstrating that mature cells can be reprogrammed to a pluripotent state. In place of ESCs and their ethical concerns, iPSCs can provide an unlimited supply of human cells and tissue for research. They can be used to study disease in vitro and test new drugs to treat the disease in a patient-specific manner. The iPSCs can also be differentiated into a specific cell type for transplantation or infusion into a patient. While a few examples have been given here, there are a whole host of possibilities for iPSCs in medicine. Cardiovascular disease, liver disease, kidney failure, diabetes and organ transplants are just a few of the areas in which iPSC technology is being used to study potential cures. We are now beginning to see the benefits of what was only dreamed when the first iPSCs were reported.

DEFINITIONS

Cellular reprogramming The ability to alter the fate of a cell by changing the genes it expresses.

Chimeric An organism whose cells have at least two distinct genotypes. One way of producing chimeric mice is to add iPSCs to a blastocyst.

Chromatin A tightly condensed complex of DNA and histone proteins.

Dedifferentiation The loss of specialized cellular characteristics; to revert back to an earlier stage of development.

Endonuclease An enzyme that cleaves the phosphodiester bond between two DNA nucleotides typically at a specific sequence unless it is a nonspecific endonuclease.

Epigenetics The regulation of gene transcription by elements around the DNA but not of the DNA. Elements that produce "open" chromatin, histone modifications, and the interaction of transcription factors with enhancers and promoters are epigenetic in nature.

Extrachromosomal DNA that is located outside of the chromosomes. In the case of viral DNA, it remains in the cytoplasm where it can replicate independently.

Induced pluripotent stem cells A pluripotent stem cell artificially derived from somatic cells by inducing the expression of specific genes.

Nucleosome The smallest unit of chromatin consisting of DNA wrapped two times (~147 base pairs) around a core of eight proteins called histones.

Solenoid A unit of chromatin consisting of six nucleosomes that form a hexagon-like structure that contributes to the compact nature of DNA within the cell's nucleus.

Transdifferentiation The regulation of transcription factors to reprogram a cell from one fate to another, specifically involving a lineage conversion. For example, the conversion of a mesodermal cell to an ectodermal cell.

Transduce The transfer of genetic material from one organism to another.

Vector A vehicle, such as a virus, used to transfer genetic material to a target cell.

References

Aasen, T., Raya, A., Barrero, M.J., Garreta, E., Consiglio, A., Gonzalez, F., Vassena, R., Bilić, J., Pekarik, V., Tiscornia, G., et al., 2008. Efficient and rapid generation of induced pluripotent stem cells from human keratinocytes. Nat. Biotechnol. 26, 1276–1284.

Alvarez, C.V., Garcia-Lavandeira, M., Garcia-Rendueles, M.E.R., Diaz-Rodriguez, E., Garcia-Rendueles, A.R., Perez-Romero, S., Vila, T.V., Rodrigues, J.S., Lear, P.V., Bravo, S.B., 2012. Defining stem cell types: understanding the therapeutic potential of ESCs, ASCs, and iPS cells. J. Mol. Endocrinol. 49, R89–R111.

Baylin, S.B., Jones, P.A., 2011. A decade of exploring the cancer epigenome—biological and translational implications. Nat. Rev. Cancer 11, 726–734.

Black, A.R., Black, J.D., Azizkhan-Clifford, J., 2001. Sp1 and krüppel-like factor family of transcription factors in cell growth regulation and cancer. J. Cell. Physiol. 188, 143–160.

Blaese, R.M., Culver, K.W., Miller, A.D., Carter, C.S., Fleisher, T., Clerici, M., Shearer, G., Chang, L., Chiang, Y., Tolstoshev, P., et al., 1995. T lymphocyte-directed gene therapy for ADA-SCID: initial trial results after 4 years. Science 270, 475–480.

Blasco, M.A., Serrano, M., Fernandez-Capetillo, O., 2011. Genomic instability in iPS: time for a break: Genomic instability in iPS. EMBO J. 30, 991–993.

Boiani, M., Schöler, H.R., 2005. Developmental cell biology: regulatory networks in embryo-derived pluripotent stem cells. Nat. Rev. Mol. Cell Biol. 6, 872–881.

Bouard, D., Alazard-Dany, N., Cosset, F.-L., 2009. Viral vectors: from virology to transgene expression. Br. J. Pharmacol. 157, 153–165.

Brehove, M., Wang, T., North, J., Luo, Y., Dreher, S.J., Shimko, J.C., Ottesen, J.J., Luger, K., Poirier, M.G., 2015. Histone Core phosphorylation regulates DNA accessibility. J. Biol. Chem. 290, 22612–22621.

Briggs, R., King, T.J., 1952. Transplantation of living nuclei from blastula cells into enucleated frogs' eggs. Proc. Natl. Acad. Sci. U. S. A. 38, 455–463.

Cyranoski, D., 2017. Japanese man is first to receive "reprogrammed" stem cells from another person. Nature 552 (7685), 302.

Daley, G.Q., Lensch, M.W., Jaenisch, R., Meissner, A., Plath, K., Yamanaka, S., 2009. Broader implications of defining standards for the pluripotency of iPSCs. Cell Stem Cell 4, 200–201. (author reply 202).

Dang, C.V., 2012. MYC on the path to cancer. Cell 149, 22–35.

Davis, R.L., Weintraub, H., Lassar, A.B., 1987. Expression of a single transfected cDNA converts fibroblasts to myoblasts. Cell 51, 987–1000.

Daya, S., Berns, K.I., 2008. Gene therapy using Adeno-associated virus vectors. Clin. Microbiol. Rev. 21, 583–593.

DePolo, N.J., Reed, J.D., Sheridan, P.L., Townsend, K., Sauter, S.L., Jolly, D.J., Dubensky, T.W., 2000. VSV-G pseudotyped lentiviral vector particles produced in human cells are inactivated by human serum. Mol. Ther. 2, 218–222.

Edelstein, M., 2017. Vectors used in gene therapy clinical trails, Gene Therapy Clinical Trials Worldwide. http://www.abedia.com/wiley/vectors.php. Accessed on July 10, 2018.

Eminli, S., Utikal, J., Arnold, K., Jaenisch, R., Hochedlinger, K., 2008. Reprogramming of neural progenitor cells into induced pluripotent stem cells in the absence of exogenous Sox2 expression. Stem Cells Dayt. Ohio 26, 2467–2474.

Fairchild, P.J., Robertson, N.J., Cartland, S., Nolan, K.F., Waldmann, H., 2005. Cell replacement therapy and the evasion of destructive immunity. Stem Cell Rev. 1 (2), 159–167.

Gurdon, J.B., 1962. The developmental capacity of nuclei taken from intestinal epithelium cells of feeding tadpoles. J. Embryol. Exp. Morphol. 10, 622–640.

Gurdon, J.B., 2013. The egg and the nucleus: A battle for supremacy (Nobel lecture). Angew. Chem. Int. Ed. 52, 13890–13899.

Gurdon, J.B., Byrne, J.A., 2003. The first half-century of nuclear transplantation. Proc. Natl. Acad. Sci. 100, 8048–8052.

Harui, A., Suzuki, S., Kochanek, S., Mitani, K., 1999. Frequency and stability of chromosomal integration of adenovirus vectors. J. Virol. 73, 6141–6146.

Hawkins, K., 2014. Cell signalling pathways underlying induced pluripotent stem cell reprogramming. World J. Stem Cells 6, 620.

Ho, L., Crabtree, G.R., 2010. Chromatin remodelling during development. Nature 463, 474–484.

Hochedlinger, K., Jaenisch, R., 2015. Induced pluripotency and epigenetic reprogramming. Cold Spring Harb. Perspect. Biol. 7.

Hohenstein Elliott, K.A., Peterson, C., Soundararajan, A., Kan, N., Nelson, B., Spiering, S., Mercola, M., Bright, G.R., 2012. Laser-based propagation of human iPS and ES cells generates reproducible cultures with enhanced differentiation potential. Stem Cells Int. 2012, 926463.

Hu, K., 2014. Vectorology and factor delivery in induced pluripotent stem cell reprogramming. Stem Cells Dev. 23, 1301–1315.

Huangfu, D., Osafune, K., Maehr, R., Guo, W., Eijkelenboom, A., Chen, S., Muhlestein, W., Melton, D.A., 2008. Induction of pluripotent stem cells from primary human fibroblasts with only Oct4 and Sox2. Nat. Biotechnol. 26, 1269–1275.

Jenuwein, T., Allis, C., 2001. Translating the histone code. Science 293, 1074–1080.

Jin, J., Jing, W., Lei, X.-X., Feng, C., Peng, S., Boris-Lawrie, K., Huang, Y., 2011. Evidence that Lin28 stimulates translation by recruiting RNA helicase A to polysomes. Nucleic Acids Res. 39, 3724–3734.

Jin, Z.-B., Okamoto, S., Xiang, P., Takahashi, M., 2012. Integration-free induced pluripotent stem cells derived from retinitis pigmentosa patient for disease modeling. Stem Cells Transl. Med. 1, 503–509.

Kamao, H., Mandai, M., Okamoto, S., Sakai, N., Suga, A., Sugita, S., Kiryu, J., Takahashi, M., 2014. Characterization of human induced pluripotent stem cell-derived retinal pigment epithelium cell sheets aiming for clinical application. Stem Cell Rep. 2, 205–218.

Kashyap, V., Rezende, N.C., Scotland, K.B., Shaffer, S.M., Persson, J.L., Gudas, L.J., Mongan, N.P., 2009. Regulation of stem cell pluripotency and differentiation involves a mutual regulatory circuit of the Nanog, OCT4, and SOX2 pluripotency transcription factors with Polycomb repressive complexes and stem cell microRNAs. Stem Cells Dev. 18, 1093–1108.

Kikuchi, T., Morizane, A., Doi, D., Magotani, H., Onoe, H., Hayashi, T., Mizuma, H., Takara, S., Takahashi, R., Inoue, H., et al., 2017. Human iPS cell-derived dopaminergic neurons function in a primate Parkinson's disease model. Nature 548, 592–596.

Kotterman, M.A., Chalberg, T.W., Schaffer, D.V., 2015. Viral vectors for gene therapy: Translational and clinical outlook. Annu. Rev. Biomed. Eng. 17, 63–89.

Lai, C.M., Lai, Y.K.Y., Rakoczy, P.E., 2002. Adenovirus and adeno-associated virus vectors. DNA Cell Biol. 21, 895–913.

Levine, M., Tjian, R., 2003. Transcription regulation and animal diversity. Nature 424, 147–151.

Loh, Y.-H., Wu, Q., Chew, J.-L., Vega, V.B., Zhang, W., Chen, X., Bourque, G., George, J., Leong, B., Liu, J., et al., 2006. The Oct4 and Nanog transcription network regulates pluripotency in mouse embryonic stem cells. Nat. Genet. 38, 431–440.

Mandai, M., Watanabe, A., Kurimoto, Y., Hirami, Y., Morinaga, C., Daimon, T., Fujihara, M., Akimaru, H., Sakai, N., Shibata, Y., et al., 2017. Autologous induced stem-cell-derived retinal cells for macular degeneration. N. Engl. J. Med. 376, 1038–1046.

Masaki, T., Qu, J., Cholewa-Waclaw, J., Burr, K., Raaum, R., Rambukkana, A., 2013. Reprogramming adult Schwann cells to stem cell-like cells by leprosy bacilli promotes dissemination of infection. Cell 152, 51–67.

Masui, S., Nakatake, Y., Toyooka, Y., Shimosato, D., Yagi, R., Takahashi, K., Okochi, H., Okuda, A., Matoba, R., Sharov, A.A., et al., 2007. Pluripotency governed by Sox2 via regulation of Oct3/4 expression in mouse embryonic stem cells. Nat. Cell Biol. 9, 625–635.

McClellan, J.S., Dove, C., Gentles, A.J., Ryan, C.E., Majeti, R., 2015. Reprogramming of primary human Philadelphia chromosome-positive B cell acute lymphoblastic leukemia cells into nonleukemic macrophages. Proc. Natl. Acad. Sci. U. S. A. 112, 4074–4079.

Mfopou, J.K., Bouwens, L., 2015. β cells at last: diabetes. Nat. Rev. Endocrinol. 11, 5–6.

Mitsui, K., Tokuzawa, Y., Itoh, H., Segawa, K., Murakami, M., Takahashi, K., Maruyama, M., Maeda, M., Yamanaka, S., 2003. The homeoprotein Nanog is required for maintenance of pluripotency in mouse epiblast and ES cells. Cell 113, 631–642.

Montini, E., Cesana, D., Schmidt, M., Sanvito, F., Bartholomae, C.C., Ranzani, M., Benedicenti, F., Sergi, L.S., Ambrosi, A., Ponzoni, M., et al., 2009. The genotoxic potential of retroviral vectors is strongly modulated by vector design and integration site selection in a mouse model of HSC gene therapy. J. Clin. Invest. 119, 964–975.

Murugan, V., 2009. Embryonic stem cell research: a decade of debate from Bush to Obama. Yale J. Biol. Med. 82 (3), 101–103.

Nakagawa, M., Koyanagi, M., Tanabe, K., Takahashi, K., Ichisaka, T., Aoi, T., Okita, K., Mochiduki, Y., Takizawa, N., Yamanaka, S., 2008. Generation of induced pluripotent stem cells without Myc from mouse and human fibroblasts. Nat. Biotechnol. 26, 101–106.

Nicholas, C.R., Kriegstein, A.R., 2010. Regenerative medicine: cell reprogramming gets direct. Nature 463, 1031–1032.

NIH Stem Cell Information Home Page, 2016. Stem Cell Information [World Wide Web site]. National Institutes of Health, U.S. Department of Health and Human Services, Bethesda, MD. Available from: https://stemcells.nih.gov/info/2001report/chapter2.htm. (accessed 05.01.18).

Nisole, S., Saib, A., 2004. Early steps of retrovirus replicative cycle. Retrovirology 1.

Niwa, H., Miyazaki, J., Smith, A.G., 2000. Quantitative expression of Oct-3/4 defines differentiation, dedifferentiation or self-renewal of ES cells. Nat. Genet. 24, 372–376.

Okita, K., Ichisaka, T., Yamanaka, S., 2007. Generation of germline-competent induced pluripotent stem cells. Nature 448, 313–317.

Pagliuca, F.W., Millman, J.R., Gürtler, M., Segel, M., Van Dervort, A., Ryu, J.H., Peterson, Q.P., Greiner, D., Melton, D.A., 2014. Generation of functional human pancreatic β cells in vitro. Cell 159, 428–439.

Pedersen, R.A., 1994. Studies of in vitro differentiation with embryonic stem cells. Reprod. Fertil. Dev. 6, 543–552.

Pesonen, S., Kangasniemi, L., Hemminki, A., 2011. Oncolytic adenoviruses for the treatment of human cancer: focus on translational and clinical data. Mol. Pharm. 8, 12–28.

Raper, S.E., Yudkoff, M., Chirmule, N., Gao, G.-P., Nunes, F., Haskal, Z.J., Furth, E.E., Propert, K.J., Robinson, M.B., Magosin, S., et al., 2002. A pilot study of in vivo liver-directed gene transfer with an adenoviral vector in partial ornithine transcarbamylase deficiency. Hum. Gene Ther. 13, 163–175.

Raper, S.E., Chirmule, N., Lee, F.S., Wivel, N.A., Bagg, A., Gao, G., Wilson, J.M., Batshaw, M.L., 2003. Fatal systemic inflammatory response syndrome in a ornithine transcarbamylase deficient patient following adenoviral gene transfer. Mol. Genet. Metab. 80, 148–158.

Rowland, B.D., Bernards, R., Peeper, D.S., 2005. The KLF4 tumour suppressor is a transcriptional repressor of p53 that acts as a context-dependent oncogene. Nat. Cell Biol. 7, 1074–1082.

Ruiz, S., Brennand, K., Panopoulos, A.D., Herrerías, A., Gage, F.H., Izpisua-Belmonte, J.C., 2010. High-efficient generation of induced pluripotent stem cells from human astrocytes. PLoS One 5, e15526.

Rybak, A., Fuchs, H., Smirnova, L., Brandt, C., Pohl, E.E., Nitsch, R., Wulczyn, F.G., 2008. A feedback loop comprising lin-28 and let-7 controls pre-let-7 maturation during neural stem-cell commitment. Nat. Cell Biol. 10, 987–993.

Schambach, A., Galla, M., Maetzig, T., Loew, R., Baum, C., 2007. Improving transcriptional termination of self-inactivating gamma-retroviral and lentiviral vectors. Mol. Ther. J. Am. Soc. Gene Ther. 15, 1167–1173.

Schepers, G.E., Teasdale, R.D., Koopman, P., 2002. Twenty pairs of sox: extent, homology, and nomenclature of the mouse and human sox transcription factor gene families. Dev. Cell 3, 167–170.

Schneuwly, S., Klemenz, R., Gehring, W.J., 1987. Redesigning the body plan of Drosophila by ectopic expression of the homoeotic gene Antennapedia. Nature 325, 816–818.

Silva, J., Nichols, J., Theunissen, T.W., Guo, G., van Oosten, A.L., Barrandon, O., Wray, J., Yamanaka, S., Chambers, I., Smith, A., 2009. Nanog is the gateway to the pluripotent ground state. Cell 138, 722–737.

Simara, P., Motl, J.A., Kaufman, D.S., 2013. Pluripotent stem cells and gene therapy. Transl. Res. 161, 284–292.

Spemann, H., 1962. Embryonic Development and Induction. Hafner Publishing Company, New York.

Sridharan, R., Tchieu, J., Mason, M.J., Yachechko, R., Kuoy, E., Horvath, S., Zhou, Q., Plath, K., 2009. Role of the murine reprogramming factors in the induction of pluripotency. Cell 136, 364–377.

Stilwell, J.L., Samulski, R.J., 2004. Role of viral vectors and virion shells in cellular gene expression. Mol. Ther. 9, 337–346.

Sumi, T., Tsuneyoshi, N., Nakatsuji, N., Suemori, H., 2007. Apoptosis and differentiation of human embryonic stem cells induced by sustained activation of c-Myc. Oncogene 26, 5564–5576.

Sun, L.T., Yamaguchi, S., Hirano, K., Ichisaka, T., Kuroda, T., Tada, T., 2014. Nanog coregulated by Nodal/Smad2 and Oct4 is required for pluripotency in developing mouse epiblast. Dev. Biol. 392, 182–192.

Takahashi, K., Yamanaka, S., 2006. Induction of pluripotent stem cells from mouse embryonic and adult fibroblast cultures by defined factors. Cell 126, 663–676.

Takahashi, K., Tanabe, K., Ohnuki, M., Narita, M., Ichisaka, T., Tomoda, K., Yamanaka, S., 2007. Induction of pluripotent stem cells from adult human fibroblasts by defined factors. Cell 131, 861–872.

Tessarz, P., Kouzarides, T., 2014. Histone core modifications regulating nucleosome structure and dynamics. Nat. Rev. Mol. Cell Biol. 15, 703–708.

Tetreault, M.-P., Yang, Y., Katz, J.P., 2013. Krüppel-like factors in cancer. Nat. Rev. Cancer 13, 701–713.

Timp, W., Feinberg, A.P., 2013. Cancer as a dysregulated epigenome allowing cellular growth advantage at the expense of the host. Nat. Rev. Cancer 13, 497–510.

Utikal, J., Maherali, N., Kulalert, W., Hochedlinger, K., 2009. Sox2 is dispensable for the reprogramming of melanocytes and melanoma cells into induced pluripotent stem cells. J. Cell Sci. 122, 3502–3510.

Varlakhanova, N.V., Cotterman, R.F., deVries, W.N., Morgan, J., Donahue, L.R., Murray, S., Knowles, B.B., Knoepfler, P.S., 2010. myc maintains embryonic stem cell pluripotency and self-renewal. Differ. Res. Biol. Divers. 80, 9–19.

Viswanathan, S.R., Powers, J.T., Einhorn, W., Hoshida, Y., Ng, T.L., Toffanin, S., O'Sullivan, M., Lu, J., Phillips, L.A., Lockhart, V.L., et al., 2009. Lin28 promotes transformation and is associated with advanced human malignancies. Nat. Genet. 41, 843–848.

Voevodin, A.F., Marx, P.A., 2009. Simian Virology. Wiley-Blackwell, Ames, IA.

Voss, T.C., Hager, G.L., 2013. Dynamic regulation of transcriptional states by chromatin and transcription factors. Nat. Rev. Genet. 15, 69–81.

Waddington, C., 1957. The Strategy of the Genes. Volume George Allen and Unwin, London, UK.

Weismann, A., 1892. Das Keimplasma: Eine Theorie der Vererbung. Gustav Fischer, Jena.

Wernig, M., Meissner, A., Cassady, J.P., Jaenisch, R., 2008. c-Myc is dispensable for direct reprogramming of mouse fibroblasts. Cell Stem Cell 2, 10–12.

Wilson, J.M., 2009. Lessons learned from the gene therapy trial for ornithine transcarbamylase deficiency. Mol. Genet. Metab. 96, 151–157.

Xu, Y.Z., Kanagaratham, C., Radzioch, D., 2013. Chromatin remodelling during host-bacterial pathogen interaction. In: Radzioch, D. (Ed.), Chromatin Remodelling. InTech.

Yan, R.-T., Li, X., Huang, J., Guidry, C., Wang, S.-Z., 2013. Photoreceptor-like cells from reprogramming cultured mammalian RPE cells. Mol. Vis. 19, 1178–1187.

Yu, J., Vodyanik, M.A., Smuga-Otto, K., Antosiewicz-Bourget, J., Frane, J.L., Tian, S., Nie, J., Jonsdottir, G.A., Ruotti, V., Stewart, R., et al., 2007. Induced pluripotent stem cell lines derived from human somatic cells. Science 318, 1917–1920.

Zapata-Linares, N., Rodriguez, S., Mazo, M., Abizanda, G., Andreu, E.J., Barajas, M., Prosper, F., Rodriguez-Madoz, J.R., 2016. Generation and characterization of human iPSC line generated from mesenchymal stem cells derived from adipose tissue. Stem Cell Res. 16, 20–23.
Zhang, S., 2014. Sox2, a key factor in the regulation of pluripotency and neural differentiation. World J. Stem Cells 6, 305.
Zhang, Y., Bergelson, J.M., 2005. Adenovirus Receptors. J. Virol. 79, 12125–12131.
Zhang, J., Ratanasirintrawoot, S., Chandrasekaran, S., Wu, Z., Ficarro, S.B., Yu, C., Ross, C.A., Cacchiarelli, D., Xia, Q., Seligson, M., et al., 2016. LIN28 regulates stem cell metabolism and conversion to primed pluripotency. Cell Stem Cell 19, 66–80.

Further Reading

Chun, Y.S., Byun, K., Lee, B., 2011. Induced pluripotent stem cells and personalized medicine: current progress and future perspectives. Anat. Cell Biol. 44, 245.
Morata, G., Sánchez-Herrero, E., 1998. Pulling the fly's leg: developmental biology. Nature 392, 657–658.

SECTION II

ADULT NEUROGENESIS

CHAPTER

5

Introduction to Adult Neurogenesis

OUTLINE

LEARNING OBJECTIVES

1. Describe the development of the subventricular zone.
2. Describe how the dentate gyrus is populated with neural stem cells during development.
3. Discuss the research contributions that led to the acceptance of adult neurogenesis.
4. Explain how proliferation markers label dividing cells and how those cells are visualized.

https://doi.org/10.1016/B978-0-12-811014-0.00005-6

INTRODUCTION

As a respected authority on neurons, Santiago Ramon y Cajal (Spanish Neurologist, 1852–1934) wrote, "In the adult centers, the nerve paths are something fixed, ended, and immutable. Everything may die, nothing may be regenerated. It is for the science of the future to change, if possible, this harsh decree" (Colucci-D'Amato et al., 2006). His words became the prevailing dogma, that no new neurons are produced in the adult brain. Thus, up until the early 1990s, neurogenesis was thought to happen only during development. Now we know that neurogenesis occurs in two distinct regions of the adult brain, the subventricular zone (SVZ) lining the lateral ventricles and the subgranular zone (SGZ) of the dentate gyrus in the hippocampus (Fig. 5.1). The niche of each region has a unique morphology, though both contain neural stem cells (NSCs) and glia, with extensive vascularization.

In the SVZ, NSCs or Type B cells are adjacent to ependymal cells that line the lateral ventricles. Type B cells give rise to Type C *neural progenitor cells* (NPCs). The NPCs produce neuroblasts (Type A) that migrate approximately 5mm through the rostral migratory stream (RMS), a narrow path defined by a scaffold of astrocytes and blood vessels, to reach the olfactory bulb where they become interneurons.

Type 1 NSCs in the SGZ produce NPCs that undergo two stages of differentiation (Type 2a/b) based on molecular expression, though both are neural-committed intermediate progenitor cells. A third type of neural-committed progenitor cell (Type 3) is unique due to its loss of Nestin expression. The Type 3 cell is a migratory neuroblast that migrates a short distance within the SGZ of the dentate gyrus and develops the morphology of an immature neuron. The extensive dendritic branches of an immature neuron form dendritic spines that promote its integration into the existing circuitry, and generate synaptic properties of a mature neuron in the dentate gyrus. More details are provided in the following chapters on the proliferation and differentiation of these cells. Here, the neurodevelopment of each niche and embryonic origin of the NSCs of each region is described.

NERVOUS SYSTEM DEVELOPMENT

Following the blastocyst stage of embryonic development, gastrulation (Chapter 2) produces three germ layers, each containing committed progenitor cells that give rise to cell types specific to that germ layer. The nervous system arises from the outermost germ layer of the developing mammalian embryo called the ectoderm. A portion of the dorsal ectoderm extending between the anterior and posterior poles differentiates into the neuroectoderm. The cells of the neuroectoderm undergo rapid cell division during

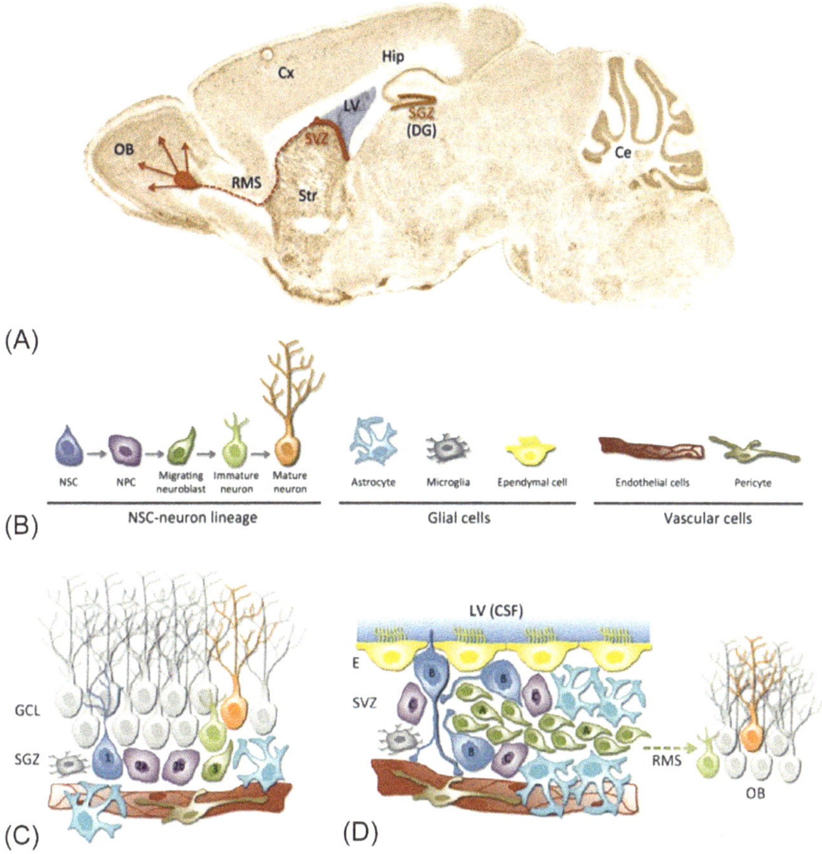

FIG. 5.1 Neurogenic niches in the adult mammalian brain. (A) Schematic representation of the neurogenic niches in a sagittal section of the adult mouse brain: the SGZ (orange) in dentate gyrus (DG) of the hippocampus (Hip), and the SVZ (red) in the lateral wall of the lateral ventricles (LV). SVZ-derived newborn neurons migrate toward the olfactory bulb (OB) through the RMS. (B) Cellular components of neurogenic niches. In addition to the NSC-neuron lineage, neurogenic niches are composed of glial cells (astrocytes, microglia, and ependymal cells) and vascular cells (endothelial cells and pericytes). (C, D) Illustration of SGZ and SVZ neurogenic niches. The cytoarchitecture and relationships between cellular components of the niche are represented. Different cell types (color and shape) correspond to those depicted in B). Note the close proximity between blood vessels and NSCs and (NPCs in both niches. C) The SGZ neurogenic niche. Radial Type 1 cells correspond to the NSCs that give rise to Type 2a/b NPCs, which differentiate into Type 3 neuroblasts. Neuroblasts migrate guided by astrocytes and become immature neurons that finally mature and integrate into the granular cell layer (GCL). D) SVZ neurogenic niche. This niche is located underneath the ependymal lining (E) of the LV. It is composed of Type B quiescent cells (NSCs), which can activate and generate Type C NPCs that rapidly proliferate and generate Type A neuroblasts. Neuroblasts migrate long distances through the RMS to the OB where they mature into interneurons. Note that a tunnel of astrocytes and a scaffold of blood vessels guide migration of neuroblasts. Also note that monociliated Type B cells can directly contact cerebrospinal fluid (CSF) and blood vessels. *Ce*, cerebelum; *Cx*, cortex; *Str*, striatum. *Image reprinted under the Creative Commons license. Copyright © 2016 Bátiz, Castro, Burgos, Velásquez, Muñoz, Lafourcade, Troncoso-Escudero and Wyneken.*

primary neurulation, producing the first recognizable ectodermal structure, the neural tube, which arises via proliferation and cell shape changes. The lateral edges of the neuroectoderm build up while the center invaginates until the lateral edges come together to form the dorsal neural tube (Fig. 5.2). The inside of the neural tube was originally the outside of the embryo. The anatomical labels refer to the embryonic origin; thus closest to the inner

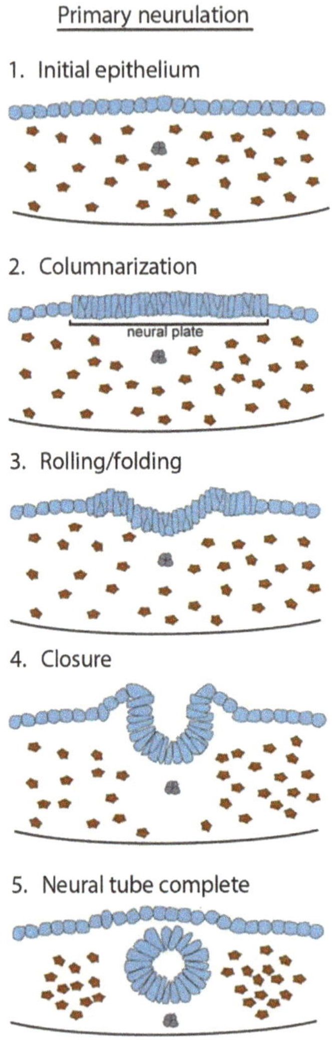

FIG. 5.2 Primary neurulation. (1) A cross section of the embryo after gastrulation is completed. Changes in (2) cell shape and increased proliferation induce (3) tissue folding and invaginations that account for (4, 5) the formation of the neural tube. *Image reprinted with permission from Lowery, L.A., Sive, H., 2004. Strategies of vertebrate neurulation and a re-evaluation of teleost neural tube formation. Mech. Dev. 121, 1189–1197.*

neural tube or lumen is the apical side or ventricular surface. When the neural tube fails to close completely the result is a neural tube defect called spina bifida. While typically not fatal, there may be nerve damage depending on whether there is an open or closed lesion resulting in disabilities such as bladder or bowel problems and weak leg muscles or scoliosis.

The cells of the neural tube are called neuroepithelial progenitors, and they produce a large number of *neural cells* over the course of development. Neuroepithelial progenitors can be identified by their expression of Nestin, an intermediate filament and marker of NSCs (Lendahl et al., 1990). At the end of primary neurulation, neuroepithelial progenitors divide symmetrically to increase their numbers (Fig. 5.3). In mouse, around

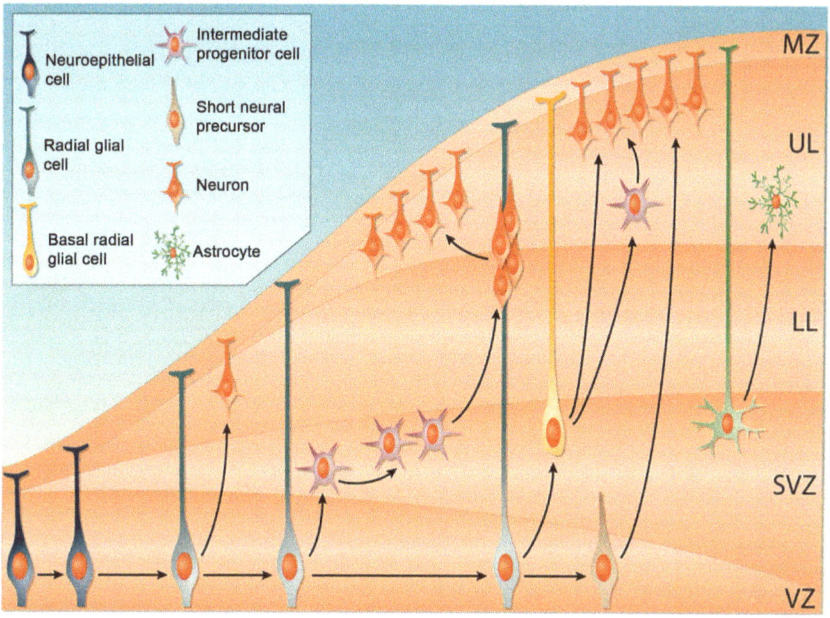

FIG. 5.3 Neural cell production in the developing cortex. Left to right) Early in neocortical development, neuroepithelial cells divide symmetrically to expand the progenitor pool before transforming into RGCs. RGCs typically divide asymmetrically to self-renew and produce either neurons or intermediate NPC. NPCs divide symmetrically to generate pairs of neurons, or in some cases additional NPCs that then make neurons. RGCs also give rise to short neural precursors and basal RGCs. Short neural precursors maintain an apical end foot and are located in the ventricular zone; they undergo terminal symmetric division to make neurons. Basal RGCs have a basal attachment at the pial basement membrane similar to RGCs, but do not maintain an apical process and thus have their cell bodies located in the outer margins of the SVZ. Basal RGCs self-renew (not shown) and generate NPCs and neurons. At the end of neurogenesis, most RGCs and basal RGCs transform into astrocyte progenitors (green cell) during the gliogenic period of embryonic development. *VZ*, ventricular zone; *SVZ*, subventricular zone; *LL*, lower layers; *UL*, upper layers; *MZ*, marginal zone. *Image reprinted with permission from Franco, S.J., Müller, U., 2013. Shaping our minds: stem and progenitor cell diversity in the mammalian neocortex. Neuron 77, 19–34.*

embryonic day nine the neuroepithelial progenitor cells begin to divide asymmetrically producing one neuron and one basal progenitor cell that will divide within the putative SVZ (Frederiksen and McKay, 1988). Shortly thereafter the neuroepithelial progenitors transform into *radial glial cells* (RGCs) that will produce cells of the cerebral cortex (Franco and Müller, 2013). RGCs are fate-restricted neurogenic cells that express glial markers such as brain lipid binding protein (BLBP) and glial fibrillary acidic protein (GFAP), but retain the neuroepithelial marker Nestin (Malatesta et al., 2003). RGCs have a cell body that remains close to the ventricular surface during proliferation, but extend a process to the pial (basal) surface that is maintained during mitosis (Mission et al., 1991). When RGCs divide symmetrically the basal process is inherited by one of the daughter cells, while the other cell produces its own process that extends to the basal surface (Miyata et al., 2001; Saito et al., 2003). However, RGCs usually divide asymmetrically, giving rise to NPCs, some short neural precursors and some basal RGCs (Franco and Müller, 2013).

Embryonic Origin of SVZ Neural Stem Cells

Although RGCs can produce glial cells, they generally give rise to neurons and are responsible for all projection neurons in the cerebral cortex—either directly or through the NPCs they produce (Anthony et al., 2004; Malatesta et al., 2003). The neurons that make up the inner or deep layers of the cortex (layers 5 and 6) are born before the neurons in the outer or upper layers (layers 2–4). New neurons use the basal process of an RGC to migrate toward the pial surface past early-born neurons (Nowakowski and Rakic, 1979; Rakic, 1971, 1972). This migratory movement is called inside-out development. Early studies suggested RGCs simply mediated the migration of committed neurons to their final position in the developing cortex, but were later shown to be the source of the neurons. Further, a subset of RGCs become adult NSCs (Fuentealba et al., 2015). Lineage analysis showed that ventricular RGCs retract the basal process and reside in the SVZ of the dorsolateral ventricles, becoming the Type B cells in the adult SVZ (Doetsch et al., 1999; Merkle et al., 2004). In rodents, these Type B cells are present throughout life and produce cells that eventually become the mature interneurons of the olfactory bulb (Chapter 7).

In 1989 Dr. Sally Temple demonstrated that in vitro, single cells isolated from the septal region of the early developing rat forebrain produce neurons (70% of clones) or glia (8%) or both neurons and glia (22%; Temple, 1989). Other labs soon showed that cells isolated from various embryonic ages and regions of the brain contained NPCs (Cattaneo and McKay, 1990; Kilpatrick and Bartlett, 1993; Reynolds et al., 1992). Subsequent studies demonstrated how factors such as fibroblast growth factor or conditioned media from immortalized astrocytes increased proliferation of NPCs

(Drago et al., 1991a, 1991b; Kilpatrick and Bartlett, 1993). The NPCs in the embryonic central nervous system (CNS) were identified as the multipotent neuroepithelial progenitors and RGCs described earlier (Rakic, 2003). These findings opened new avenues of research that led to a greater understanding of neurodevelopment, an important platform for NSC research.

Embryonic Origin of Hippocampal Neural Stem Cells

The rodent hippocampus looks like a backwards C-shaped structure in the caudal brain (Fig. 5.4A). The top of the C is called the septal pole, located medial and caudal to the lateral ventricle. The bottom end is called the temporal pole that ends caudal to the amygdala. Adult neurogenesis occurs in a region of the hippocampus called the dentate gyrus. The dentate gyrus of the hippocampus consists of the granule cell layer and molecular layer (Fig. 5.1B). Cell bodies in the granule cell layer of the dentate gyrus extend dendrites that radiate toward the molecular layer and an axon toward the hilus, the region just proximal to the dentate gyrus.

Development of the dentate gyrus occurs in two phases, beginning in the most ventral temporal pole along the lateral ventricle (Li et al., 2013). At embryonic day 15.5 NSCs initially occupy the ventricular zone near the temporal pole and migrate the short distance (~100 μm) to the putative dentate gyrus. Through postnatal day zero, NSCs migrate dorsally throughout the hippocampus toward the septal pole, filling in the rest of the dentate gyrus (Ahn and Joyner, 2005; Bai et al., 2002; Li et al., 2009, 2013). Similar to cortical development, the processes of RGCs extend from the

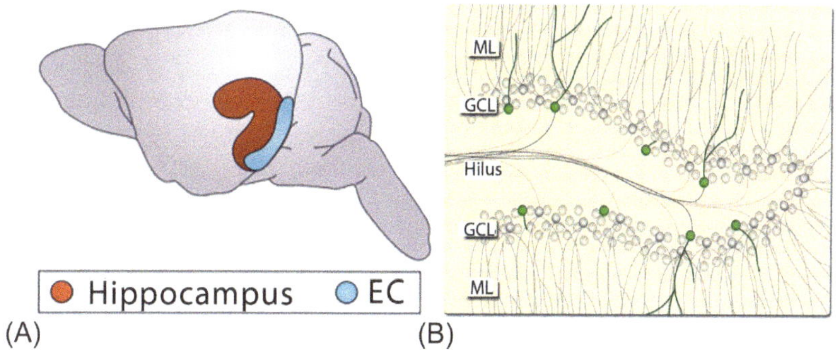

FIG. 5.4 The rodent hippocampus. (A) The hippocampus (red) is a large structure in close proximity to the entorhinal cortex (blue). (B) A cross section through the hippocampus. Mature granule cell neuron cell bodies reside in the GCL and extend dendrites into the molecular layer (ML), while the axon extends through the hilus. The cell bodies of immature neurons (green) arise in the SGZ. *(A) Reprinted with permission from Strange, B.A., Witter, M.P., Lein, E.S., Moser, E.I., 2014. Functional organization of the hippocampal longitudinal axis. Nat. Rev. Neurosci. 15, 655–669; (B) reprinted under the Creative Commons license Copyright © 2011 Rosenzweig and Wojtowicz.*

ventricular zone to the pial surface forming the boundaries of the dentate gyrus (Nowakowski, 1991). Neurons produced in the ventricular zone migrate along the RGC processes, becoming progressively more differentiated and complex along the way (Nowakowski and Rakic, 1979). The NSCs are maintained in the SGZ and produce new neurons throughout life that are thought to contribute to the refinement of network function.

DISCOVERY AND ACCEPTANCE OF ADULT NEUROGENESIS

It took several decades of research before adult neurogenesis was widely accepted by neuroscientists. Several areas of research converged to provide evidence that demonstrated the birth of new neurons in the adult brain. Credit is given to Joseph Altman and Gopal D. Das for the first observation of neurogenesis in the adult rat hippocampus in 1965 (Altman and Das, 1965). Tritiated thymidine ([3H]thymidine) incorporates into the DNA of dividing cells and the radioactive nucleotides can be visualized using autoradiography. Rats injected with [3H]thymidine were sacrificed at time points ranging from 6h to 20days later (Box 5.2). Labeled cells in the dentate gyrus could be clearly seen in what Altman coined the SGZ of the dentate gyrus. A similar study was carried out 2 years later in guinea pig, and in 1969 images of [3H]thymidine-labeled cells in the rat forebrain and olfactory bulb were published, though they were largely ignored by the scientific community (Altman, 1969; Altman and Das, 1967). However, as the plastic nature of the nervous system was discerned, the concept of adult neurogenesis became more accepted.

Evidence of adult neurogenesis was published again in Kaplan and Hinds (1977). They also injected adult rats with [3H]thymidine and after 30days, examined the brains using electron microscopy of the autoradiograph images. They reported labeled cells in the dentate gyrus of the hippocampus and the olfactory bulb. Although Hinds was an expert on the olfactory bulb, there was much skepticism regarding the identity of the labeled cells as neurons, a crucial criterion upon which the idea of adult neurogenesis hinged. A study published by Yale University's Pasko Rakic in 1985 proffered the idea that adult neurogenesis was phylogenetically restricted to lower vertebrates such as rodents (Rakic, 1985). Further, Rakic reported no evidence of adult neurogenesis in a nonhuman primate. The Rakic paper put a damper on adult neurogenesis studies in higher vertebrates for ~20years.

Neural Plasticity in the Adult Brain

Support for adult neurogenesis continued to be found in other lower vertebrate species. First, a study on the canary *Serinus canaria* showed that a vocal control nucleus in the telencephalon contained adult-generated neurons (Paton and Nottebohm, 1984). Adult birds were treated with [3H] thymidine; 30 days later responses to auditory stimuli were elicited from neurons in the nucleus *hyperstriatum ventralis pars caudalis* (HVc) using an electrode filled with horseradish peroxidase. The dye-injected cells were also labeled with [3H]thymidine. This study was important in that it demonstrated functional neuronal properties in cells born in the adult brain. The same lab showed that the birth of new neurons in nucleus HVc is tied to the development of a new song repertoire every year in April (springtime mating) followed by the loss of many neurons by October (Kirn and Nottebohm, 1993; Kirn et al., 1994). Neurogenesis in birds was more readily accepted, but their potential relevance to mammals was not generally acknowledged (Nottebohm, 1985).

Another vertebrate study in the adult teleost fish *Sternarchus albifrons* demonstrated the presence of neurons able to incorporate [3H]thymidine (Anderson and Waxman, 1985). While it was known that the caudal tail tip of the fish retained a regenerative ability, the authors showed that explant cultures from the rostral spinal cord of an adult teleost were also capable of neurogenesis. The [3H]thymidine-labeled cells were confirmed to be neurons both morphologically and by labeling with neuron-specific filament antibodies. While this study provided more support for adult neurogenesis, it failed, as with songbirds to provide relevance to higher mammals.

The pervasive theory of adaptability or plasticity postulated that plasticity and adaptability could be accomplished by "strengthening synapses" without requiring structural reorganization (Hebb, 1949). Leading up to the paradigm shift of adult neurogenesis, evidence accumulated in support of neuronal plasticity. In the 1960s it was known that axonal sprouting after injury occurred in the peripheral nervous system. Later studies showed axonal elongation and synaptic reorganization in the CNS in response to injury (Lynch et al., 1973; Moore et al., 1971; Raisman, 1969). A study by Greenough et al. (1978) found that enriched or social environments promoted structural changes now known to be associated with synaptic strengthening at the pre- and postsynapse (Box 5.1).

One of the current leaders in the field of adult neurogenesis, Fred Gage, began studying fetal rat tissue grafts in normal, aging and diseased rat brains in the mid-1980s. His mentor, Anders Björklund had established grafting methods that Gage and colleagues used to publish a number of studies demonstrating the ability of fetal tissue such as monoaminergic or cholinergic neurons of the striatum, basal forebrain, and mesencephalon or septum to incorporate into adult brain regions (Brundin et al., 1985; Gage et al., 1985a, b; Isacson et al., 1985; Segal et al., 1985). They found a high rate of survival and evidence of migration (Dunnett et al., 1984). Several reports found that the grafted neurons established connections and became fully

BOX 5.1

SPOTLIGHT ON ACTION POTENTIALS

The electrical activity of neurons was studied in the squid by Sir Alan Hodgkin (1914–98) and Sir Andrew Huxley (1917–2012). Some squid have axons large enough to see with the naked eye, one millimeter in diameter. By piercing the axon with a microelectrode they could measure the voltage inside the axon at rest, called the resting *membrane potential*, and when it fired an *action potential*, a pulse of increased voltage that travels rapidly through the axon to the axon terminal (Hodgkin et al., 1952). Through experimental manipulation they discovered the basis of an action potential (Fig. 5.5).

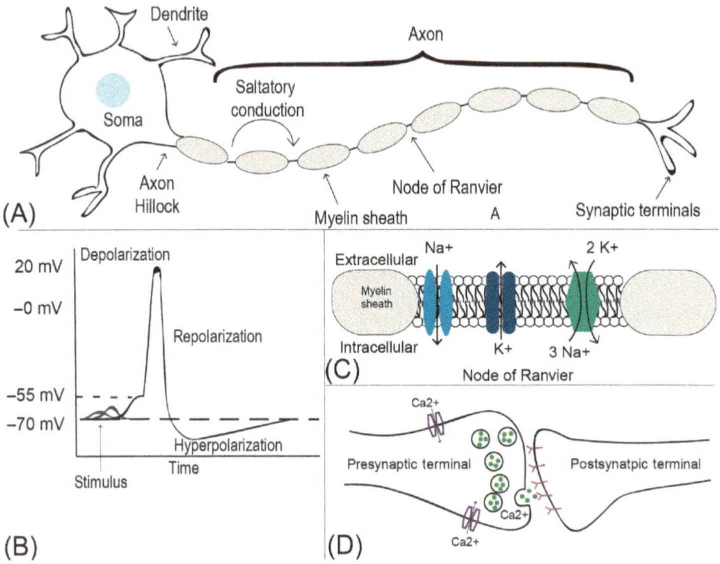

FIG. 5.5 The action potential. (A) Anatomy of a neuron. (B) The change in membrane potential over time starting at rest (−70 mV). A stimulus causes the membrane to depolarize, but must reach the threshold of excitation (−55 mV) before an action potential will occur. (C) A node of Ranvier between the myelin sheaths. When the membrane is depolarized voltage-gated Na^+ channels allow Na^+ into the cell and voltage-gated K^+ channels allow K^+ to leave the cell. The sodium-potassium pump returns the ions back to their original position by exchanging three Na^+ ions for two K^+ ions. (D) The synapse. In the presynaptic terminal, the depolarizing action potential opens Ca^{2+} channels. The Ca^{2+} results in synaptic vesicles filled with neurotransmitter to fuse with the membrane, releasing its contents into the synaptic cleft. Depending on the type of neurotransmitter, binding to postsynaptic receptors results in either an excitatory (depolarizing) or inhibitory (hyperpolarizing) change in the postsynaptic potential due to the opening of Na^+, Ca^{2+}, or Cl^- channels, respectively.

BOX 5.1

SPOTLIGHT ON ACTION POTENTIALS—CONT'D

In order to understand how an action potential occurs, it is first necessary to describe the forces that are active when a neuron is at rest. *Electrostatic pressure* is one of the forces that affects a cell's membrane potential, measured in volts. Electrostatic pressure is the attraction of opposite charges and the repulsion of like charges. The potential of a neuron is determined by the presence of charged particles like ions, such as sodium (Na^+), potassium, (K^+), chloride (Cl^-), and proteins, which have an overall negative charge. When a neuron is at rest the proteins inside contribute to a negative resting potential, on average between −60 and −70 mV, depending on the neuron. Neurons have a high concentration of K^+ inside the cell and are surrounded externally by a high concentration of Na^+. The Na^+ outside the cell is attracted to the negative charge inside the cell and is closely opposed to the lipid membrane separating the charges.

Voltage-gated ion channels are proteins embedded in the neuron's membrane that open and close based on the voltage inside the cell. Voltage-gated Na^+ channels are closed when the cell is at rest and open when the membrane potential (voltage) becomes more positive than −55 mV. When voltage-gated Na^+ channels open, the second force affecting the membrane potential comes into play—*diffusion*. Diffusion is the movement of molecules from a region of high concentration to regions of low concentration. The high concentration of Na^+ on the outside rushes into the neuron, a region of low Na^+ concentration. Hodgkin and Huxley found that by injecting small amounts of positive current, they could cause the neuron to fire an action potential. The potential at which enough voltage-gated Na^+ channels open to start an action potential is called the *threshold of excitation*. The movement of Na^+ into the neuron causes a *depolarization*—a positive change in membrane potential. During an action potential the voltage reaches as high as +20 mV. This change is only temporary due to voltage-gated K^+ channels also present in the membrane that open when the membrane potential becomes more positive than 0 mV. The high concentration of K^+ inside the cell diffuses out through the voltage-gated K^+ channels to a region of low concentration, returning the neuron to its resting membrane potential (Hodgkin and Huxley, 1952). A more negative change in membrane potential is called a *hyperpolarization*.

Continued

BOX 5.1

SPOTLIGHT ON ACTION POTENTIALS—CONT'D

In order for ion concentrations to be returned to their original concentration and location, a pump is necessary to transfer them back across the membrane. The sodium-potassium pump uses ATP as energy to transfer three Na^+ ions out of the cell while transferring two K^+ ions into the cell (Clausen et al., 2017). The pump is always active to maintain or restore the ion concentrations in anticipation of another action potential.

The process of membrane depolarization is dynamic and happens on a rapid time scale. Voltage-gated sodium channels are located around the neuron's soma but are highly concentrated at a region called the axon hillock, the base of the axon as it protrudes from the cell. When the threshold of excitation is reached, the amount of Na^+ that enters at the axon hillock is such that diffusion forces the ions down the axon toward the synaptic terminal. The axon is wrapped in myelin, which acts as an insulator to prevent the exchange of ions. At regularly spaced intervals nodes of Ranvier are small gaps between the myelin where a concentration of voltage-gated Na^+ and K^+ channels are opened due to the Na^+ diffusing down the axon. The depolarization of the axon "jumps" from node to node, a movement called saltatory conduction, until the depolarization reaches the synaptic terminal.

When the synaptic terminal is depolarized it releases *neurotransmitter*. Voltage-gated Ca^{2+} channels in the membrane of the synaptic terminal open due to the depolarization by Na^+ influx. The Cas^{2+} causes synaptic vesicles filled with neurotransmitter to fuse with the membrane, releasing the contents into the gap between the axon terminal and the postsynaptic terminal on the adjacent dendrite. This gap is called the synaptic cleft, an extraordinarily narrow space that neurotransmitters cross and bind to receptors, causing the opening of neurotransmitter- (or ligand-)gated ion channels to open. The result is an excitatory postsynaptic potential (EPSP) if the ligand-gated channel is a pore for positively charged ions like Na^+ or Ca^{2+}. Glutamate, for example is an excitatory neurotransmitter and its release causes an EPSP. GABA is a neurotransmitter that only gates Cl^- channels and the opening causes the negatively charged Cl^- to diffuse into the neuron, resulting in hyperpolarization, called an inhibitory postsynaptic potential (IPSP). This is how neurons communicate excitation and inhibition across neural networks.

integrated in the adult brain (Björklund et al., 1983; Freund et al., 1985; Gage and Björklund, 1986; Segal et al., 1985). Further, the grafted neurons were able to reverse chemical lesion-induced Parkinson disease-like symptoms (Dunnett et al., 1981; Freed et al., 1980; Perlow et al., 1979). These studies supported the idea of plasticity in the adult brain.

Advances in immunohistochemical markers for proliferating neurons finally allowed researchers to show concrete evidence of adult neurogenesis in the mammalian brain. Antibodies that labeled newly generated neurons targeted neuronal proteins such as cell division cycle 2 (Cdc2) and neural cell adhesion molecule (NCAM). Antibody staining was correlated with [3H]thymidine expression in the rat dentate gyrus, demonstrating that the new cells were neuronal (Okano et al., 1993; Seki and Arai, 1993). Next, the introduction of the thymidine analog 5-bromo-2′-deoxyuridine (BrdU) provided an immunohistochemical tool that allowed confocal microscopic visualization of cells double labeled with neuronal marker fluorescence and BrdU fluorescence (Box 5.2) (Kuhn et al., 1996; Miller and Nowakowski, 1988; Seki and Arai, 1995).

BOX 5.2

BIRTH DATING PROLIFERATING CELLS

The discovery of NSCs in the adult brain led investigators to ask (1) how frequently do NSCs divided, (2) what kind of cells arise from adult NSCs, and (3) what is the final fate of adult-generated cells in the brain. Autoradiography has been used since 1958 to visualize dividing cells and the technique was applied to studying adult neurogenesis (Altman and Das, 1965; Firket and Verly, 1958). An autoradiograph uses an X-ray film or nuclear emulsion plate to record charged particles. Dividing cells are labeled by injecting a living animal with thymidine conjugated to tritium, (written as [3H]), a radioactive isotope of hydrogen. [3H]thymidine is incorporated into the DNA of dividing cells, becoming a permanent part of the cell. If the animal is sacrificed immediately after the injection, the [3H] thymidine indicates which cells were dividing at the time of injection. If the animal is allowed to live days, weeks or months after the injection, the [3H]thymidine indicates the fate of cells that were dividing at the time of injection. Ultrathin sections of tissue are on an X-ray film or nuclear emulsion plate. Cells in the tissue that have the [3H]thymidine in its DNA release tritium decay particles onto the X-ray or emulsion plate. The Alvarez-Buylla lab used a nuclear liquid emulsion in which decay particles from the [3H]thymidine exposed silver grains, causing them to

Continued

BOX 5.2

BIRTH DATING PROLIFERATING CELLS—CONT'D

turn black (Fig. 5.6) (Doetsch et al., 1999). Afterwards, the exposed silver grains over nuclei containing [3H]thymidine were visualized using electron microscopy. The technique was used to demonstrate that Type B cells gave rise to Type C cells (Doetsch et al., 1999).

The thymidine analog 5-bromo-2′-deoxyuridine (BrdU) is used to label dividing cells. BrdU is injected intraperitoneally and incorporates into DNA being replicated where it will remain for the life of the cell. If one wants to observe the fate of adult-born cells, animals injected with BrdU can be sacrificed at various time points such as days, weeks or months after the injection. The brain tissue can be processed for immunohistochemistry using an antibody that targets BrdU, revealing the location of cells that divided at the time of injection. For example, an animal sacrificed 2 weeks after BrdU injection will have BrdU-positive Type B and/or C cells in the subventricular zone (SVZ) and Type A neuroblasts in the olfactory bulb that will become interneurons. BrdU has been an important tool, providing evidence of adult neurogenesis, final fate and the function of cells born in the adult brain (Gonzalez-Perez et al., 2009; Kornack and Rakic, 1999, 2001; Zhao et al., 2014).

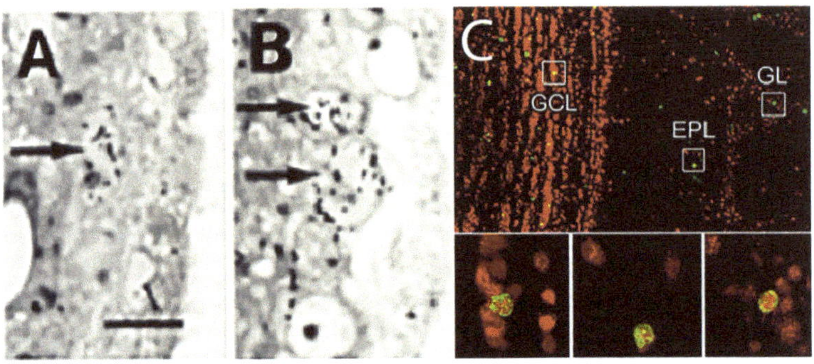

FIG. 5.6 Markers of proliferating cells. (A, B) Mice were injected with [3H]thymidine that is seen in thin sections of the SVZ imaged with electron microscopy. Dividing cells show silver grains of [3H]thymidine overlying the nucleus (arrows). Scale bar = 10 μm. (C) Mice were injected with BrdU and sacrificed 4 weeks later. BrdU (green) colocalizes with the neuronal nuclear protein marker NeuN (red) in the GCL, external plexiform layer (EPL) and glomerular layer (GL) of the olfactory bulb.(A, B) Reprinted with permission from Doetsch, F., Caillé, I., Lim, D.A., García-Verdugo, J.M., Alvarez-Buylla, A., 1999. Subventricular zone astrocytes are neural stem cells in the adult mammalian brain. Cell 97, 703–716. (C) Modified under the Creative Commons license Copyright © 2016 Snyder et al.

Following the discovery of NPCs in embryonic brain by Temple and others, researchers prominent in the field of embryonic neural development repeated experiments that isolated neural cells in culture, this time obtained from adult rodent brain between 3 and 18 months old (Reynolds and Weiss, 1992; Temple and Raff, 1985). They found that while the majority of isolated cells died, a small fraction proliferated, producing both neurons and glia. Based on an understanding of cortical development, scientists focused on the SVZ and found proliferating and migratory cells in the adult brain of rodents (Lois and Alvarez-Buylla, 1994; Luskin, 1993; Morshead et al., 1994).

In the late 1990s adult neurogenesis was finally demonstrated in primates. Dr. Elizabeth Gould demonstrated in 1998 that BrdU-positive cells colocalized with the neuronal marker Neuron Specific Enolase (NSE) in the dentate gyrus of adult marmosets, a New World primate (Gould et al., 1998). A year later she published another paper showing BrdU-positive neurons in the dentate gyrus of adult macaque monkeys, an Old World primate more closely related to humans (Gould et al., 1999; Han et al., 2007). Another strong piece of evidence came from Peter S. Eriksson who injected consenting terminally ill cancer patients with BrdU. The deceased brains were processed for immunohistochemistry to reveal the survival of adult generated neurons ranging from 16 days postinjection to more than 2 years old in patients whose ages ranged from 57 to 72 years at the time of death (Eriksson et al., 1998). The idea that new neurons are only generated in lower vertebrates was put to rest.

The following chapters cover the two neurogenic regions of the adult brain, the SVZ in the dorsolateral ventricles and the dentate gyrus of the hippocampus. While commonalities are found between cells of the two regions, the niche cells, neural stem cell morphology, migration and final fate of the adult-born neurons have many unique characteristics. The final chapter explores some of the areas of research in which adult neurogenesis has been found to play a role in or correlate with states of health and disease. It examines whether the promise of stem cells is also true of NSCs in the adult brain.

DEFINITIONS

Action potential A pulse of increased voltage that travels rapidly through a neuron.

Depolarization A more positive change in potential between the inside and outside of the plasma membrane of a nerve cell due to migration of sodium ions to the intracellular side.

Diffusion The movement of molecules from a region of high concentration to regions of low concentration.

Electrostatic pressure The attractive force between charged particles with opposite signs or the repulsive force between two charged particles with the same sign.

Hyperpolarization A more negative change in potential between the inside and outside of the plasma membrane of a nerve cell due to migration of potassium ions to the extracellular side.

Membrane potential The voltage measured inside a cell relative to the voltage just outside it.

Neural cells Encompasses neuronal and glial cells.

Neural progenitor cell (NPC) Similar to transit-amplifying cells (Chapter 2). These are cells that proliferate for a finite period of time, giving rise to and then differentiating into neuronal or glial cell types.

Neurotransmitter A chemical released from the synaptic terminal that diffuses across the synaptic cleft that confers an excitatory or inhibitory potential to the postsynaptic neuron.

Primary neurulation A process in vertebrate embryos in which a segment of the developing ectoderm transforms into neuroectoderm or neural plate, followed by a series of proliferation and folding that results in the neural tube.

Radial glial cell (RGC) Radially oriented, bipolar cells in the developing vertebrate cortex that have a basal process that persists during cell division to support the migration of newborn neurons. RGCs express astrocytic markers but retain expression of the neuroepithelial marker, intermediate filament Nestin. RGCs give rise to neurons, oligodendrocytes and astrocytes, and are the precursor of Type B astrocytes in the SVZ.

Threshold of excitation The potential at which a depolarization must reach for an action potential to occur.

References

Ahn, S., Joyner, A.L., 2005. In vivo analysis of quiescent adult neural stem cells responding to Sonic hedgehog. Nature 437, 894–897.

Altman, J., 1969. Autoradiographic and histological studies of postnatal neurogenesis. IV. Cell proliferation and migration in the anterior forebrain, with special reference to persisting neurogenesis in the olfactory bulb. J. Comp. Neurol. 137, 433–457.

Altman, J., Das, G.D., 1965. Autoradiographic and histological evidence of postnatal hippocampal neurogenesis in rats. J. Comp. Neurol. 124, 319–335.

Altman, J., Das, G.D., 1967. Postnatal neurogenesis in the guinea-pig. Nature 214, 1098–1101.

Anderson, M.J., Waxman, S.G., 1985. Neurogenesis in tissue cultures of adult teleost spinal cord. Brain Res. 352, 203–212.

Anthony, T.E., Klein, C., Fishell, G., Heintz, N., 2004. Radial glia serve as neuronal progenitors in all regions of the central nervous system. Neuron 41, 881–890.

Bai, C.B., Auerbach, W., Lee, J.S., Stephen, D., Joyner, A.L., 2002. Gli2, but not Gli1, is required for initial Shh signaling and ectopic activation of the Shh pathway. Development (Camb., Engl.) 129, 4753–4761.

Björklund, A., Stenevi, U., Schmidt, R.H., Dunnett, S.B., Gage, F.H., 1983. Intracerebral grafting of neuronal cell suspensions. I. Introduction and general methods of preparation. Acta Physiol. Scand. Suppl. 522, 1–7.

Brundin, P., Barbin, G., Isacson, O., Mallat, M., Chamak, B., Prochiantz, A., Gage, F.H., Björklund, A., 1985. Survival of intracerebrally grafted rat dopamine neurons previously cultured in vitro. Neurosci. Lett. 61, 79–84.

Cattaneo, E., McKay, R., 1990. Proliferation and differentiation of neuronal stem cells regulated by nerve growth factor. Nature 347, 762–765.

Clausen, M.V., Hilbers, F., Poulsen, H., 2017. The structure and function of the Na,K-ATPase isoforms in health and disease. Front Physiol 8, 371.

Colucci-D'Amato, L., Bonavita, V., di Porzio, U., 2006. The end of the central dogma of neurobiology: stem cells and neurogenesis in adult CNS. Neurol. Sci. 27, 266–270.

Doetsch, F., Caillé, I., Lim, D.A., García-Verdugo, J.M., Alvarez-Buylla, A., 1999. Subventricular zone astrocytes are neural stem cells in the adult mammalian brain. Cell 97, 703–716.

Drago, J., Murphy, M., Bailey, K.A., Bartlett, P.F., 1991a. A method for the isolation of purified murine neuroepithelial cells from the developing mouse brain. J. Neurosci. Methods 37, 251–256.

Drago, J., Murphy, M., Carroll, S.M., Harvey, R.P., Bartlett, P.F., 1991b. Fibroblast growth factor-mediated proliferation of central nervous system precursors depends on endogenous production of insulin-like growth factor I. Proc. Natl. Acad. Sci. U. S. A. 88, 2199–2203.

Dunnett, S.B., Björklund, A., Stenevi, U., Iversen, S.D., 1981. Grafts of embryonic substantia nigra reinnervating the ventrolateral striatum ameliorate sensorimotor impairments and akinesia in rats with 6-OHDA lesions of the nigrostriatal pathway. Brain Res. 229, 209–217.

Dunnett, S.B., Bunch, S.T., Gage, F.H., Björklund, A., 1984. Dopamine-rich transplants in rats with 6-OHDA lesions of the ventral tegmental area. I. Effects on spontaneous and drug-induced locomotor activity. Behav. Brain Res. 13, 71–82.

Eriksson, P.S., Perfilieva, E., Björk-Eriksson, T., Alborn, A.-M., Nordborg, C., Peterson, D.A., Gage, F.H., 1998. Neurogenesis in the adult human hippocampus. Nat. Med. 4, 1313–1317.

Firket, H., Verly, W.G., 1958. Autoradiographic visualization of synthesis of deoxyribonucleic acid in tissue culture with tritium-labelled thymidine. Nature 181, 274–275.

Franco, S.J., Müller, U., 2013. Shaping our minds: stem and progenitor cell diversity in the mammalian neocortex. Neuron 77, 19–34.

Frederiksen, K., McKay, R., 1988. Proliferation and differentiation of rat neuroepithelial precursor cells in vivo. J. Neurosci. 8, 1144–1151.

Freed, W.J., Perlow, M.J., Karoum, F., Seiger, A., Olson, L., Hoffer, B.J., Wyatt, R.J., 1980. Restoration of dopaminergic function by grafting of fetal rat substantia nigra to the caudate nucleus: long-term behavioral, biochemical, and histochemical studies. Ann. Neurol. 8, 510–519.

Freund, T.F., Bolam, J.P., Björklund, A., Stenevi, U., Dunnett, S.B., Powell, J.F., Smith, A.D., 1985. Efferent synaptic connections of grafted dopaminergic neurons reinnervating the host neostriatum: a tyrosine hydroxylase immunocytochemical study. J. Neurosci. 5, 603–616.

Fuentealba, L.C., Rompani, S.B., Parraguez, J.I., Obernier, K., Romero, R., Cepko, C.L., Alvarez-Buylla, A., 2015. Embryonic origin of postnatal neural stem cells. Cell 161, 1644–1655.

Gage, F.H., Björklund, A., 1986. Enhanced graft survival in the hippocampus following selective denervation. Neuroscience 17, 89–98.

Gage, F.H., Stenevi, U., Carlstedt, T., Foster, G., Björklund, A., Aguayo, A.J., 1985a. Anatomical and functional consequences of grafting mesencephalic neurons into a peripheral nerve "bridge" connected to the denervated striatum. Exp. Brain Res. 60, 584–589.

Gage, F.H., Brundin, P., Isacson, O., Björklund, A., 1985b. Rat fetal brain tissue grafts survive and innervate host brain following five day pregraft tissue storage. Neurosci. Lett. 60, 133–137.

Gonzalez-Perez, O., Romero-Rodriguez, R., Soriano-Navarro, M., Garcia-Verdugo, J.M., Alvarez-Buylla, A., 2009. Epidermal growth factor induces the progeny of subventricular zone Type B cells to migrate and differentiate into oligodendrocytes. Stem Cells 27, 2032–2043.

Gould, E., Tanapat, P., McEwen, B.S., Flugge, G., Fuchs, E., 1998. Proliferation of granule cell precursors in the dentate gyrus of adult monkeys is diminished by stress. Proc. Natl. Acad. Sci. 95, 3168–3171.

Gould, E., Reeves, A.J., Fallah, M., Tanapat, P., Gross, C.G., Fuchs, E., 1999. Hippocampal neurogenesis in adult Old World primates. Proc. Natl. Acad. Sci. 96, 5263–5267.

Greenough, W.T., West, R.W., DeVoogd, T.J., 1978. Subsynaptic plate perforations: changes with age and experience in the rat. Science 202, 1096–1098.

Han, K., Konkel, M.K., Xing, J., Wang, H., Lee, J., Meyer, T.J., Huang, C.T., Sandifer, E., Hebert, K., Barnes, E.W., et al., 2007. Mobile DNA in Old World monkeys: a glimpse through the rhesus macaque genome. Science 316, 238–240.

Hebb, D., 1949. The Organization of Behavior. A Neuropsychological Theory. Wiley, New York, NY.

Hodgkin, A.L., Huxley, A.F., 1952. Movement of sodium and potassium ions during nervous activity. Cold Spring Harb. Symp. Quant. Biol. 17, 43–52.

Hodgkin, A.L., Huxley, A.F., Katz, B., 1952. Measurement of current-voltage relations in the membrane of the giant axon of Loligo. J. Physiol. 116 (4), 424–448.

Isacson, O., Brundin, P., Gage, F.H., Björklund, A., 1985. Neural grafting in a rat model of Huntington's disease: progressive neurochemical changes after neostriatal ibotenate lesions and striatal tissue grafting. Neuroscience 16, 799–817.

Kaplan, M.S., Hinds, J.W., 1977. Neurogenesis in the adult rat: electron microscopic analysis of light radioautographs. Science 197, 1092–1094.

Kilpatrick, T.J., Bartlett, P.F., 1993. Cloning and growth of multipotential neural precursors: requirements for proliferation and differentiation. Neuron 10, 255–265.

Kirn, J.R., Nottebohm, F., 1993. Direct evidence for loss and replacement of projection neurons in adult canary brain. J. Neurosci. 13, 1654–1663.

Kirn, J., O'Loughlin, B., Kasparian, S., Nottebohm, F., 1994. Cell death and neuronal recruitment in the high vocal center of adult male canaries are temporally related to changes in song. Proc. Natl. Acad. Sci. U. S. A. 91, 7844–7848.

Kornack, D.R., Rakic, P., 1999. Continuation of neurogenesis in the hippocampus of the adult macaque monkey. Proc. Natl. Acad. Sci. U. S. A. 96, 5768–5773.

Kornack, D.R., Rakic, P., 2001. The generation, migration, and differentiation of olfactory neurons in the adult primate brain. Proc. Natl. Acad. Sci. U. S. A. 98, 4752–4757.

Kuhn, H.G., Dickinson-Anson, H., Gage, F.H., 1996. Neurogenesis in the dentate gyrus of the adult rat: age-related decrease of neuronal progenitor proliferation. J. Neurosci. 16, 2027–2033.

Lendahl, U., Zimmerman, L.B., McKay, R.D., 1990. CNS stem cells express a new class of intermediate filament protein. Cell 60, 585–595.

Li, G., Kataoka, H., Coughlin, S.R., Pleasure, S.J., 2009. Identification of a transient subpial neurogenic zone in the developing dentate gyrus and its regulation by Cxcl12 and reelin signaling. Development 136, 327–335.

Li, G., Fang, L., Fernández, G., Pleasure, S.J., 2013. The Ventral hippocampus is the embryonic origin for adult neural stem cells in the dentate gyrus. Neuron 78, 658–672.

Lois, C., Alvarez-Buylla, A., 1994. Long-distance neuronal migration in the adult mammalian brain. Science 264, 1145–1148.

Luskin, M.B., 1993. Restricted proliferation and migration of postnatally generated neurons derived from the forebrain subventricular zone. Neuron 11, 173–189.

Lynch, G., Deadwyler, S., Cotman, G., 1973. Postlesion axonal growth produces permanent functional connections. Science 180, 1364–1366.

Malatesta, P., Hack, M.A., Hartfuss, E., Kettenmann, H., Klinkert, W., Kirchhoff, F., Götz, M., 2003. Neuronal or glial progeny. Neuron 37, 751–764.

Merkle, F.T., Tramontin, A.D., Garcia-Verdugo, J.M., Alvarez-Buylla, A., 2004. Radial glia give rise to adult neural stem cells in the subventricular zone. Proc. Natl. Acad. Sci. 101, 17528–17532.

Miller, M.W., Nowakowski, R.S., 1988. Use of bromodeoxyuridine-immunohistochemistry to examine the proliferation, migration and time of origin of cells in the central nervous system. Brain Res. 457, 44–52.

Mission, J.-P., Austin, C.P., Takahashi, T., Cepko, C.L., Caviness, V.S., 1991. The alignment of migrating neural cells in relation to the murine neopallial radial glial fiber system. Cereb. Cortex 1, 221–229.

Miyata, T., Kawaguchi, A., Okano, H., Ogawa, M., 2001. Asymmetric inheritance of radial glial fibers by cortical neurons. Neuron 31, 727–741.

Moore, R.Y., Björklund, A., Stenevi, U., 1971. Plastic changes in the adrenergic innervation of the rat septal area in response to denervation. Brain Res. 33, 13–35.

Morshead, C.M., Reynolds, B.A., Craig, C.G., McBurney, M.W., Staines, W.A., Morassutti, D., Weiss, S., van der Kooy, D., 1994. Neural stem cells in the adult mammalian forebrain: a relatively quiescent subpopulation of subependymal cells. Neuron 13, 1071–1082.

Nottebohm, F., 1985. Neuronal replacement in adulthood. Ann. N. Y. Acad. Sci. 457, 143–161.

Nowakowski, R.S., 1991. Neuronal migration and differentiation during normal and genetically perturbed development of the hippocampal formation. In: Developmental Neuropathology of Schizophrenia. Springer Science+Business Media, New York, NY, pp. 29–60.

Nowakowski, R.S., Rakic, P., 1979. The mode of migration of neurons to the hippocampus: a Golgi and electron microscopic analysis in foetal rhesus monkey. J. Neurocytol. 8, 697–718.

Okano, H.J., Pfaff, D.W., Gibbs, R.B., 1993. RB and Cdc2 expression in brain: correlations with 3H-thymidine incorporation and neurogenesis. J. Neurosci. 13, 2930–2938.

Paton, J.A., Nottebohm, F.N., 1984. Neurons generated in the adult brain are recruited into functional circuits. Science 225, 1046–1048.

Perlow, M.J., Freed, W.J., Hoffer, B.J., Seiger, A., Olson, L., Wyatt, R.J., 1979. Brain grafts reduce motor abnormalities produced by destruction of nigrostriatal dopamine system. Science 204, 643–647.

Raisman, G., 1969. Neuronal plasticity in the septal nuclei of the adult rat. Brain Res. 14, 25–48.

Rakic, P., 1971. Neuron-glia relationship during granule cell migration in developing cerebellar cortex. A Golgi and electronmicroscopic study in Macacus rhesus. J. Comp. Neurol. 141, 283–312.

Rakic, P., 1972. Mode of cell migration to the superficial layers of fetal monkey neocortex. J. Comp. Neurol. 145, 61–83.

Rakic, P., 1985. Limits of neurogenesis in primates. Science 227, 1054–1056.

Rakic, P., 2003. Developmental and evolutionary adaptations of cortical radial glia. Cereb. Cortex (New York, NY) 1991 (13), 541–549.

Reynolds, B.A., Weiss, S., 1992. Generation of neurons and astrocytes from isolated cells of the adult mammalian central nervous system. Science 255, 1707–1710.

Reynolds, B., Tetzlaff, W., Weiss, S., 1992. A multipotent EGF-responsive striatal embryonic progenitor cell produces neurons and astrocytes. J. Neurosci. 12, 4565–4574.

Saito, K., Kawaguchi, A., Kashiwagi, S., Yasugi, S., Ogawa, M., Miyata, T., 2003. Morphological asymmetry in dividing retinal progenitor cells. Develop. Growth Differ. 45, 219–229.

Segal, M., Bjorklund, A., Gage, F.H., 1985. Transplanted septal neurons make viable cholinergic synapses with a host hippocampus. Brain Res. 336, 302–307.

Seki, T., Arai, Y., 1993. Highly polysialylated neural cell adhesion molecule (NCAM-H) is expressed by newly generated granule cells in the dentate gyrus of the adult rat. J. Neurosci. 13, 2351–2358.

Seki, T., Arai, Y., 1995. Age-related production of new granule cells in the adult dentate gyrus. Neuroreport 6, 2479–2482.

Temple, S., 1989. Division and differentiation of isolated CNS blast cells in microculture. Nature 340, 471–473.

Temple, S., Raff, M.C., 1985. Differentiation of a bipotential glial progenitor cell in a single cell microculture. Nature 313, 223–225.

Zhao, Z., Liu, H., Xiao, K., Yu, M., Cui, L., Zhu, Q., Zhao, R., Li, G.-D., Zhou, Y., 2014. Ghrelin administration enhances neurogenesis but impairs spatial learning and memory in adult mice. Neuroscience 257, 175–185.

Further Reading

Bátiz, L.F., Castro, M.A., Burgos, P.V., Velásquez, Z.D., Muñoz, R.I., Lafourcade, C.A., Troncoso-Escudero, P., Wyneken, U., 2016. Exosomes as novel regulators of adult neurogenic niches. Front. Cell. Neurosci. 9.

Lowery, L.A., Sive, H., 2004. Strategies of vertebrate neurulation and a re-evaluation of teleost neural tube formation. Mech. Dev. 121, 1189–1197.

Strange, B.A., Witter, M.P., Lein, E.S., Moser, E.I., 2014. Functional organization of the hippocampal longitudinal axis. Nat. Rev. Neurosci. 15, 655–669.

CHAPTER

6

Adult Neurogenesis in the Hippocampus

OUTLINE

LEARNING OBJECTIVES

1. Describe the anatomy of the hippocampus, including the afferent and efferent circuitry.
2. Describe the cellular events in each of the four phases of adult-born cell maturation, including morphological changes.

https://doi.org/10.1016/B978-0-12-811014-0.00006-8

3. Discuss the contribution of each cell in the niche to adult neurogenesis.
4. Explain the signaling pathways that balance Type 1 cell maintenance with quiescence.
5. List the behavioral and molecular factors that promote proliferation of Type 1 and intermediate neural progenitors.
6. Describe how the number of times an intermediate progenitor divides is regulated.
7. List the behavioral and molecular factors that influence the survival of adult-born cells.
8. Describe the ion channel activity that is observed during maturation and integration of an adult-born cell.
9. Discuss the different types of memory and which brain region(s) are involved in its retention.
10. Explain how long-term potentiation contributes to memory formation.

ANATOMY OF THE HIPPOCAMPUS

The name hippocampus is derived from two Greek words, *hippos*, meaning "horse," and *kampos*, meaning "sea monster" (Fig. 6.1). A visual comparison of the hippocampus and a seahorse reveals why this nomenclature was chosen. In humans the hippocampi are located in the temporal lobes, surrounded by the parahippocampal gyrus (Fig. 6.2). The hippocampus is involved in many aspects of learning and memory. It is essential for declarative memory (facts or episodic memories), short-term, working and spatial memory, and behavioral learning such as instrumental (operant) conditioning or classical conditioning (Box 6.1). In this chapter we will review the process of adult neurogenesis in the dentate gyrus, the cell types that contribute to its niche in the hippocampus and how adult neurogenesis is thought to play a role in learning and memory.

Afferents and Efferents of the Hippocampus

The hippocampus receives afferent information coming from many structures in the brain, including many connected through the fornix, such as the supramammillary nucleus, nucleus accumbens, medial prefrontal cortex, and the septal nuclei (Fig. 6.2). The hippocampal formation consists of the dentate gyrus, subicular cortex, and Ammon's horn, which is further divided into three regions—Cornu Ammonis 1 (CA1), 2 (CA2), and 3 (CA3; Fig. 6.4A). In the parahippocampal gyrus, the entorhinal cortex contains both afferent and efferent connections from the association cortices, the neocortex, olfactory bulb and limbic cortex (Raslau et al., 2015).

FIG. 6.1 The hippocampus and the seahorse. The hippocampus was named due to its resemblance to the seahorse by combining two Greek words, *hippos*, meaning "horse," and *kampos*, meaning "sea monster." *Image by Professor Laszlo Seress. Reprinted under the Creative Commons Attribution-Share Alike 1.0 Generic license.*

Direct afferents also arrive from the amygdala and contralateral hippocampus. The entorhinal input is referred to as the perforant path and these axons form excitatory glutamatergic synapses with granule cells of the dentate gyrus, as well as CA3 and CA1 pyramidal neurons (Deng et al., 2010). Of the six layers of the entorhinal cortex, the dentate gyrus receives input from layer II via the lateral and medial perforant path. Some afferents of layer II project directly to CA3, while CA1 receives some input from layer III via the temporoammonic pathway (Deng et al., 2010).

Dentate gyrus granule cells have dendrites in the molecular layer and an axon in the hilus, a layer proximal to the dentate gyrus. Granule cell dendrites are covered with *dendritic spines* that receive inhibitory input from local *gamma-aminobutyric acid* (GABA) interneurons in the hilus. Mossy cell bodies in the hilus extend their axon to the inner molecular layer where they provide glutamatergic input to granule cell dendrites (Fig. 6.4B; Deng et al., 2010; Scharfman and Myers, 2013).

Neurons in the hippocampus form a trisynaptic circuit. Synaptic input from the entorhinal cortex via the perforant path synapses with dentate gyrus neurons and is relayed through the mossy fiber pathway to the CA3 region by granule cell axons, or mossy fibers. The axons of CA3 pyramidal neurons are called Schaffer collaterals, which project to CA1 where they make glutamatergic synapses with CA1 pyramidal neurons, the projection

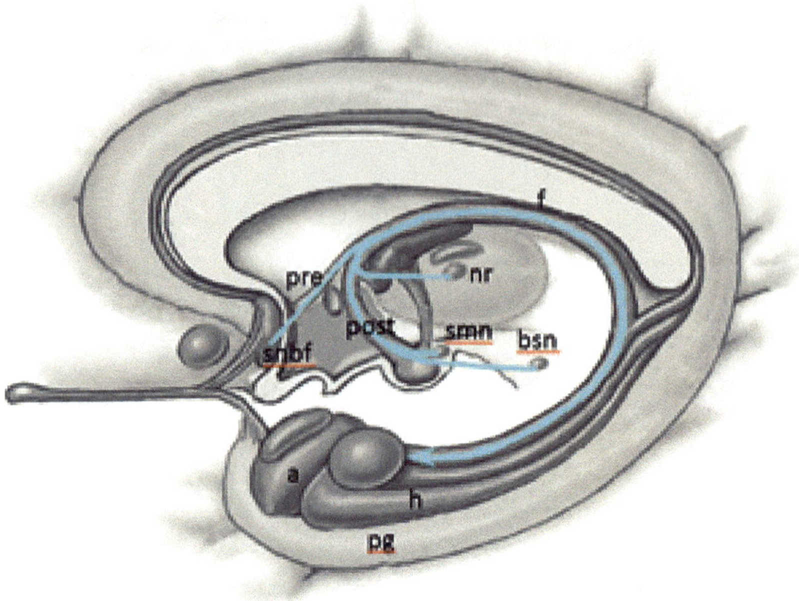

FIG. 6.2 Anatomy of the hippocampus within the limbic system and the structures that make afferent connections with the hippocampus. Abbreviations: *a*, amygdala; *bsn*, brainstem nuclei; *h*, hippocampus; *nr*, nucleus reuniens; *pg*, parahippocampal gyrus; *post*, postcommissural tracts; *pre*, precommissural tracts; *smn*, supramammillary nucleus; *snbf*, septal nuclei and basal forebrain. *Reprinted from Raslau, F.D., Mark, L.P., Sabsevitz, D.S., Ulmer, J.L., 2015. Imaging of functional and dysfunctional episodic memory. Semin. Ultrasound CT MRI 36, 260–274, with permission from Elsevier.*

neurons of the hippocampus (Owen and Grover, 2015). Efferent information projects back through the fornix and entorhinal cortex to the same areas that provided the initial input. Current research suggests that the hippocampus modifies and links afferent information in a way that allows one to form a memory that retains the relationship between the input (Eichenbaum, 2014; Komorowski et al., 2009). The interconnected nature of the hippocampus enables its involvement in memory consolidation, emotional responses, navigation and spatial orientation (Deng et al., 2010).

The hippocampus is not the storage site of long-term memories. Its role is to index certain types of memories in an appropriate cerebral area, and then to retrieve those memories when needed (Tonegawa et al., 2015). A study on patients with bilateral damage to the hippocampus suggests that it serves to retrieve more recent memories from the past few years (Manns et al., 2003). Functional imaging showed that recent memory recall activated the hippocampus with a gradual decrease over one- to nine-year-old memories, and then remained at a low constant for retrieving remote memories (10–30 years). Conversley, brain regions in the frontal

BOX 6.1

LEARNING AND MEMORY

Learning can be divided into several categories. Perceptual learning refers to the ability to recognize sensations we have experienced before. Classical conditioning is when one learns to associate an automatic, species-typical response with a stimulus that was previously unrelated to the behavior (Pavlov's dogs). Instrumental (operant) conditioning is learning an association between a response and a stimulus that involves consequences, either reinforcing stimuli that promote a particular response or punishing stimuli that discourage the response. Classical conditioning and instrumental conditioning both involve motor learning as part of the response or behavior. Indeed, instrumental conditioning can include learning a complex skill or a series of movements. Relational learning is comprehending the relationship between individual stimuli, including episodic memories. Being able to recall a sequence of events associates each event with the time that it occurred.

Memory can be divided into two broad categories: declarative memory and procedural memory. Declarative memory is the ability to remember facts, such as the current year, past experiences, your address or to distinguish between the names of items in your native language or a foreign language (Cohen and Squire, 1980). Studies of patients with amnesia due to hippocampal damage provided insight on the role of the hippocampus in learning and memory. While the hippocampus is essential for declarative memory, it is not required for procedural memory. Thus amnesic patients retain perceptual learning, stimulus-response learning and motor learning though they do not recall the learning experiences (episodic memory).

Short-term memory is typically defined as being able to store information for a short period. An example of this is parking your car while you go in to shop. You never think about where the car is parked while shopping, but upon exiting the store, you are able to retrieve the memory of where the car is located. Working memory is the ability to store and manipulate information until the opportunity to use it arrives, such as during a conversation, debate or when doing mental arithmetic. The hippocampus is essential for the process of remembering short-term information such as novel objections or associations (Jonides et al., 2008). However, studies show that the hippocampus is not the location of long- or short-term memories, but instead is required for converting short-term memories into long-term memories (Maviel, 2004). The prefrontal cortex and anterior cingulate cortex are sites of long-term memory storage (Maviel, 2004).

Continued

BOX 6.1

LEARNING AND MEMORY—CONT'D

Another type of memory dependent on the hippocampus is spatial memory. Being able to walk through a familiar room in the dark without tripping on the furniture is due to several cell types in the hippocampus and adjacent entorhinal cortex called place cells, grid cells and border cells. Place cells are neurons that encode a subset of environmental cues that correspond to a specific location (Fig. 6.3; Shapiro et al., 1997). Grid cells of the entorhinal cortex fire at multiple locations across a defined space (Moser et al., 2015). Boundary or border entorhinal cells fire along one or several borders of a defined space (Solstad et al., 2008). Drs. John M. O'Keefe, May-Britt Moser, and Edvard I. Moser have been the leaders in the field of hippocampal spatial memory and navigation since their discovery of place cells. In 2014 they were awarded the Nobel Prize in Medicine and Physiology for their work in this area of research.

FIG. 6.3 Properties of granule cells in the dentate gyrus. Left, place cells fire in response to a specific location. Middle, grid cells fire at multiple locations across a defined space. Right, border cells fire along one or more borders of a known area.

and parietal lobes showed an increase in activation when recalling memories older than 1–9 years, and greater activation when recalling remote memories (Smith and Squire, 2009). These data, as well as other studies, show that the hippocampus is important for the formation and maintenance of declarative memories for a period of a few years after learning (Manns et al., 2003; Smith and Squire, 2009).

PHASES OF ADULT-BORN CELL MATURATION

It takes approximately 7–8 weeks for an adult-born cell to become a mature neuron (Fig. 6.6; Zhao, 2006). During this time, a new cell will

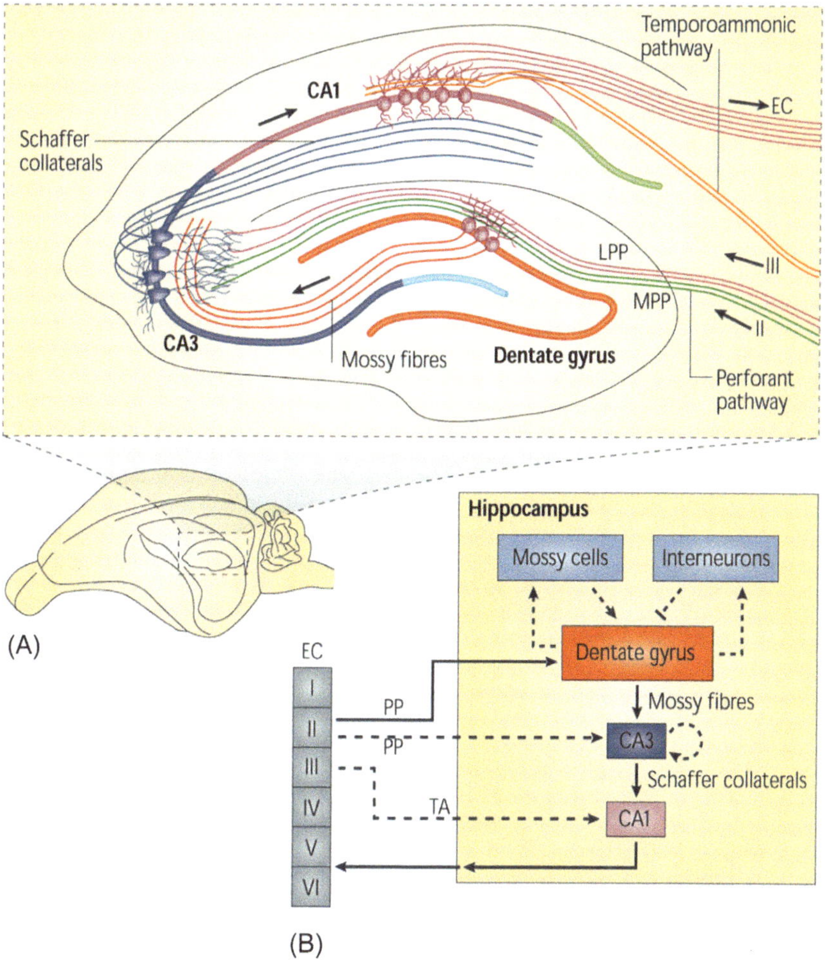

FIG. 6.4 An illustration of the afferent and efferent synapses in the hippocampus. (A) In the dentate gyrus (orange) mature granule cells (pink) have dendrites in the molecular layer. Granule cells extend axons (aka mossy fibers) through the hilus toward area CA3 (blue). Entorhinal cortex (EC) afferent (perforant path; PP) and efferent connections are shown making excitatory glutamatergic synapses with granule cells of the dentate gyrus and CA3 pyramidal neurons. Some afferents project directly to CA1 through the temporoammonic pathway (TA). *LPP*, lateral perforant pathway; *MPP*, medial perforant pathway. (B) Layers of the EC and the hippocampal neural circuitry. Solid arrows indicate the trisynaptic circuit between the EC and dentate gyrus, CA3 and CA1, and the output back through the EC. Dashed arrows indicate other connections between the different cells of the hippocampus and the EC. *Reprinted from Deng, W., Aimone, J.B., Gage, F.H., 2010. New neurons and new memories: how does adult hippocampal neurogenesis affect learning and memory? Nat. Rev. Neurosci. 11, 339–350, with permission from Nature Publishing Group.*

pass through four phases of adult neurogenesis: (1) stem cell maintenance, activation and fate specification, (2) expansion of intermediate neural progenitors, (3) migration and initial pruning of newborn granule cells, and (4) maturation and functional integration of newborn neurons (Song et al., 2012a). An adult-born cell destined to become a mature granule cell may cycle through the first two phases in the first 4–7 days, the third phase in the next 7–10 days, with the fourth phase lasting up to 4 weeks (Lucassen et al., 2010). Many adult-born cells do not survive past the third phase. The surviving cells integrate into the existing circuitry and play an important role in learning and memory.

Phase 1: Stem Cell Maintenance, Activation, and Fate Specification

At the base of the granule cell layer, Type 1 cell bodies reside in the subgranular zone (SGZ) adjacent to the hilus (Fig. 6.6, green cell). Type 1 cells have a long process that extends through the granule layer similar to radial glia cells of the developing nervous system; however, they have a dendritic arbor that branches extensively in the adjacent inner molecular layer. Lateral processes of the Type 1 cells are often in contact with other Type 1 cells. A proximal process or endfoot maintains contact with a blood vessel. A primary cilium extends into the hilus.

Some of the names used to refer to Type 1 cells include astrocyte-like cells, radial glia-like astrocytes, radial astrocytes, radial glia-like cells and adult NSCs (Huttmann et al., 2003; Kempermann, 2015; Nicola et al., 2015; Seri et al., 2001; Toda et al., 2018; Zhang and Jiao, 2015). Type 1 cells express markers of both neurons (*Nestin*) and glia (*glial fibrillary acidic protein*, GFAP; Table 6.1 for this and other markers).

In Phase 1 the Type 1 cells are typically maintained in a quiescent state by GABA signaling, but can be activated depending on experience and age to expand the pool of progenitor cells that give rise to neurons (Dranovsky and Leonardo, 2012; Encinas et al., 2006; Song et al., 2012b). During cell division, Type 1 cells can self-renew by symmetric cell division or give rise to either neurons or glia by asymmetric cell division (Song et al., 2012a).

Phase 2: Expansion of Intermediate Neural Progenitors

Type 2 and 3 cells are a population of intermediate neural progenitors that amplify the neural progenitor pool through limited self-renewal (Fig. 6.6, yellow cell; Kempermann et al., 2004). Cells may exist in this phase for approximately 2 weeks with increasingly restricted potency as they differentiate from Type 2a to Type 2b to Type 3. However, within 3 days of birth, many cells begin to express immature granule cell markers (Phase 3; Kempermann et al., 2004)

TABLE 6.1 Molecular Biomarkers That Define the Six Maturational Stages of Adult-Born Neurons

Cell type	Markers
Type 1	GFAP, Nestin, BLBP, Sox2
Type 2a	Nestin, BLBP, Sox2, Tbr2
Type 2b	Nestin, Tbr2, DCX, NeuroD, Prox-1
Type 3	Tbr2, DCX, NeuroD, Prox-1, PSA-NCAM,
Immature neuron	DCX, NeuroD, Prox-1, PSA-NCAM, Calretinin, NeuN
Mature neuron	NeuN, Calbindin

Encinas, J.M., Vaahtokari, A., Enikolopov, G., 2006. Fluoxetine targets early progenitor cells in the adult brain. Proc. Natl. Acad. Sci. 103, 8233–8238; Nicola, Z., Fabel, K., Kempermann, G., 2015. Development of the adult neurogenic niche in the hippocampus of mice. Front. Neuroanat. 9; Zhang, J., Jiao, J., 2015. Molecular biomarkers for embryonic and adult neural stem cell and neurogenesis. BioMed Res. Int. 2015, 1–14.

Type 2a cells continue to express Nestin and GFAP as well as *brain lipid binding protein* (BLBP) and *Sox2* like the Type 1 cells. It is unclear whether Type 2a cells are restricted to a neuronal or glial lineage, though they begin to express *T-box brain protein 2* (Tbr2), a transcription factor critical for neural differentiation (Arnold et al., 2008; Hodge et al., 2012). Morphologically, the loss of the radial process differentiates Type 2a cells from Type 1 cells (Kempermann et al., 2004). Instead, Type 2 cells often have horizontal processes (Fukuda et al., 2003; Steiner et al., 2006).

Type 2b cells appear to have the same role as Type 2a cells, which is to proliferate and produce neuronal-committed, intermediate progenitors. However they no longer express GFAP and now express the neuronal markers *doublecortin* (DCX), *NeuroD*, and *Prospero-related homeobox 1* (Prox-1; Hodge et al., 2008; Kempermann et al., 2004). Prox-1 promotes neuronal differentiation via Wnt signaling and is expressed in Type 2b and Type 3 cells (Karalay et al., 2011). Type 2 cells are sensitive to physiological stimuli such as physical exercise (Kronenberg et al., 2003) or pharmacological stimulation that can increase their proliferation (Encinas et al., 2006; Kronenberg et al., 2003).

Type 3 cells are neuronal-committed progenitors, apparent by their expression of DCX, *polysialic neural cell adhesion molecule* (PSA-NCAM), NeuroD and Prox-1 (Fig. 6.6, left red cell). Lack of Nestin expression distinguishes Type 3 cells from Type 2b cells. Type 3 cell processes become vertical and have less proliferative activity as they transition from an intermediate progenitor to a postmitotic immature neuron or neuroblast (Song et al., 2012a). One thicker process begins to extend toward the granule cell layer and another toward the hilus, the beginnings of dendrites and an axon, respectively (Bergami and Berninger, 2012).

Phase 3: Migration and Initial Pruning of Newborn Granule Cells

New adult-born cells migrate a short distance through the granule layer from the SGZ using the processes of radial glia (Leavitt et al., 1999; Rickmann et al., 1987; Suh et al., 2007). As the cells differentiate into neurons in the inner third of the granule cell layer, they project dendrites into the inner molecular layer, and then extend an axon toward the CA3 region (Hastings and Gould, 1999). Though the migratory distance is small, proper regulation ensures that newborn cells remain and mature within the lower third of the granule cell layer (Duan et al., 2007; Schultheiß et al., 2013). The immature neurons will spend approximately 2 weeks in this phase of postmitotic maturation.

Similar to neurodevelopment, most adult-born cells are eliminated within days after birth (Kempermann, 2003; Kempermann et al., 2004). Once new neurons exit the cell cycle, either survival or apoptotic pathways are activated. One study found that approximately 80% of adult-born neurons die before maturation and of these, the majority did not yet express the Type 2b marker DCX (Kempermann, 2003). Those cells that expressed early neuronal markers such as *Tuj1* and DCX lost approximately half their numbers over the next 7 days (Kempermann, 2003). Surviving immature granule cells are characterized by expression of the neuronal markers *calretinin* and *NeuN* (Fig. 6.6, right red cell). These cells are postmitotic and no longer self-renew. Novel experience such as environmental enrichment increases the survival of adult-born neurons. Thus, the first 2 weeks after birth is a critical period for enrichment-induced survival (Tashiro et al., 2007).

Immature granule cells are highly plastic, and long-term potentiation (LTP) is easily induced (Box 6.2). In Chapter 5, the idea of synaptic plasticity was introduced as it pertained to neuronal changes after injury and the ability of grafted neurons to integrate into adult brains (Björklund et al., 1983; Brundin et al., 1985; Lynch et al., 1973). LTP is a biological process of plasticity that strengthens the synapses in groups of nerve cells and is believed to be the neural basis of learning. Plasticity may be observed in three non-exclusive ways: changes in protein expression resulting in (1) more postsynaptic receptors, (2) more neurotransmitter release from the presynaptic terminal, and/or (3) physical changes in which the synaptic area increases (Fig. 6.7). All three of these changes cause a strengthening of the synapse.

The dendrites of immature neurons in the molecular layer are initially spineless and express both GABA and *N*-methyl-D-aspartate (NMDA)-type glutamate receptors (Box 6.2). NMDA receptors are calcium (Ca^{2+}) channels that consist of four subunits, two obligate GluN1 subunits and two variable GluN subunits (four GluN2 subunits, GluN2A-D and two

BOX 6.2

LONG-TERM POTENTIATION

Almost 100 years ago Richard Semon proposed that a memory consisted of lasting changes in the brain due to simultaneous excitations (Semon, 1921). It was almost 30 years later before Hebb was able to specify synaptically connected neurons with Semon's proposed excitations (Hebb, 1949). Tremendous progress has been made by combining behavioral experience studies and molecular mechanisms to understand how LTP underpins memory formation (Poo et al., 2016).

The basis of LTP is a strengthening of synapses by repeated stimulation (i.e., activation or excitation). Sparse, unique activity patterns of short, high-frequency bursts in CA3 produces pattern separation (Jung and McNaughton, 1993; Leutgeb et al., 2007). Pattern separation is what enables the discrete representation of a memory or *engram* (Poo et al., 2016). For example, only about 2% of granule cells are active during a hippocampus-dependent task such as exploring a new environment (Chawla et al., 2005). Researchers made small changes to an environment while observing the pattern of place cell activity in the dentate gyrus (Leutgeb et al., 2007). A cell that was strongly active in one environment showed subsequently weaker activation as the environment was slowly changed, and a new cell began to be active that associated with the new environment. The experiment demonstrated both spatial specificity and sparse activation of dentate gyrus cells.

The molecular mechanisms that underlie LTP begin and end with glutamate receptors. Glutamatergic NMDA receptors are expressed on the postsynaptic dendritic spines (Fig. 6.5). When the synapse is at rest, an ion of Mg^{2+} blocks the Ca^{2+} pore of the NMDA receptor channel, the positive charge of Mg^{2+} is attracted to the internal negative potential of the neuron. When glutamate is released from a perforant path neuron, it binds to both AMPA and NMDA receptors. AMPA receptors, which control a Na^{+} channel, open immediately and the influx of Na+ depolarizes the cell. Now the positively charged Mg^{2+} ion is repulsed, and the glutamate bound-NMDA receptor can now open, allowing an influx of Ca^{2+} ions (Lynch et al., 1983). Calcium activates the type II calcium-calmodulin kinase (CaM-KII) that ultimately results in the insertion of additional AMPA receptors, *strengthening the synapse* (Silva et al., 1992; Westmark et al., 2010; Yao et al., 2008). Eventually, there may be an increase in synaptic spines and a persistent production of new AMPA receptors as well as an increase in the size or shape of the dendritic spines or a new dendritic spine (Fig. 6.7; Bourne and Harris, 2007; Nagerl et al., 2007; Shi, 1999).

Continued

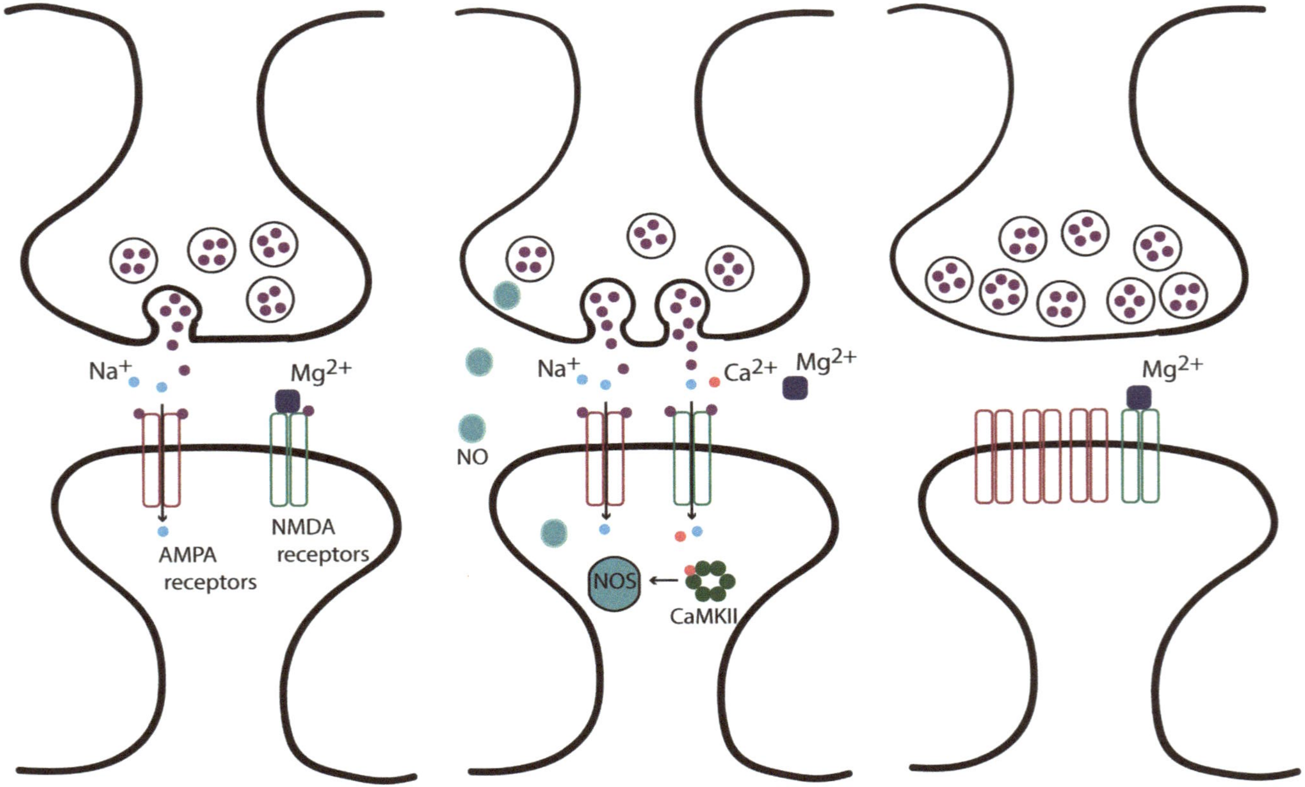

FIG. 6.5 The molecular process of LTP that strengthens the synapse. (Left) At rest, the NMDA receptor channel is blocked by Mg^{2+} ions attracted to the negative potential inside the cell. Glutamate released from the presynapse excites AMPA receptors, through which the influx of Na+ depolarizes the cell, repulsing the Mg^{2+}. (Center) The glutamate-bound NMDA receptors can now open, allowing Ca^{2+} inside the cell, which activates CaMKII. (Right) Through additional signaling, activation of CaMKII results in the insertion of more AMPA receptors.

BOX 6.2

LONG-TERM POTENTIATION—CONT'D

Strengthening of the synapse is believed to occur on the presynapse that opposes the potentiated postsynaptic terminal. Nitric oxide (NO) is a soluble gas that can easily diffuse between cells, though only over short distances and only for a brief period before it is destroyed (Hardingham et al., 2013). Neurons of the hippocampus express a CaM-KII activated enzyme called nitric oxide synthase (NOS) adjacent to the postsynaptic density that produces NO in the dendritic spines (Doucet et al., 2012; Cabrera-Pastor et al., 2016). The small amount of NO produced diffuses to the presynaptic terminal where it affects neurotransmitter release by (1) increasing synaptic vesicle release probability, (2) increasing the rate of endocytosis and vesicle recycling, (3) altering the availability and size of the readily releasable pool, and (4) increasing the expression of presynaptic proteins important for neurotransmitter release (Hardingham et al., 2013). The effects of NO are consistent with LTP-induced plasticity.

GluN3 subunits, GluN3A-B). GABAergic interneurons, such as hilar cells, first make synaptic contacts that result in tonic, depolarizing currents due to high intracellular chloride (Cl−) and the expression of the sodium-potassium-chloride (Na^+-K^+-Cl^-) cotransporter, NKCC1, by the immature neuron (Ge et al., 2006). Perforant path afferents establish contact with the new dendrites as well, releasing glutamate that cannot excite the immature neurons unless there is GABA-induced depolarization at the same time (Box 6.2; Ge et al., 2006). Coincident, GABA-mediated depolarization and NMDA receptor activation by glutamate induces a signaling cascade that results in the expression of *α-amino-3-hydroxy-5-methyl-4-isoxazolepropionic acid* (AMPA) receptors. AMPA receptors are excited by glutamate without the requirement of coincident-GABA depolarization (Ge et al., 2006). Increased AMPA receptor insertion due to neuronal excitation is one of the three types of plasticity induced by LTP (Fig. 6.7). Mature granule cells also undergo LTP, but a train of intense, high-frequency stimulation is required. Only new neurons between 4 and 6 weeks of age show a sensitivity to weak stimuli that elicits coincident, pre- and postsynaptic activity resulting in LTP (Ge et al., 2007). In experiments of environmental enrichment or exercise, immature granule cells have accelerated integration and maturation, suggesting they are an activity-dependent population. Thus the activity of immature neurons are believed to be an underlying mechanism of learning and memory (van Praag et al., 1999).

Phase 4: Maturation and Functional Integration of Newborn Neurons

When adult-born cells differentiate into mature neurons, they express NeuN, NeuroD, and Prox-1. They switch from the expression of calretinin to *calbindin* (Fig. 6.6, blue cell; Kempermann et al., 2004). Dendritic spine number increases during this time followed by spine pruning as the neurons establish connections with afferent fibers and local interneurons.

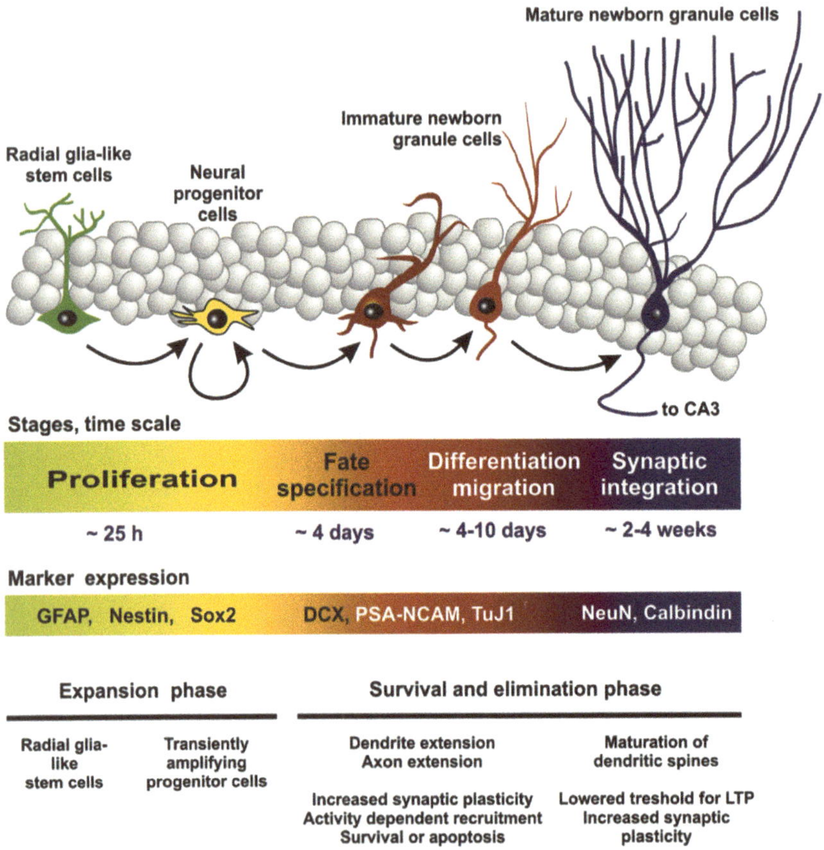

FIG. 6.6 An illustration of the phases of adult-born cell maturation in the dentate gyrus. (Top) The morphology of cells in each phase from Type 1 to a neuron, before dendritic spine development. (Middle) Approximate times a cell spends in each phase and some of the markers used to identify cells in each phase (details provided in Definitions). (Bottom) A brief description of the cells in two broad categories of adult neurogenesis: the expansion phase and survival and elimination phase. *Reprinted from Lucassen, P.J., Meerlo, P., Naylor, A.S., van Dam, A.M., Dayer, A.G., Fuchs, E., Oomen, C.A., Czéh, B., 2010. Regulation of adult neurogenesis by stress, sleep disruption, exercise and inflammation: implications for depression and antidepressant action. Eur. Neuropsychopharmacol. 20, 1–17, with permission from Elsevier.*

The fine-tuning of synaptic connections for 1–2 weeks completes the maturation of an adult-born neuron.

THE NEUROGENIC NICHE OF THE HIPPOCAMPUS

As adult-born cells proliferate, survive, and mature, they are surrounded by a number of cell types that support and/or contribute to the process of adult neurogenesis. The neurogenic niche in the adult brain is a restricted region that retains the embryonic capability to generate neurons (Song et al., 2002). The niche environment is divided into three domains (Fig. 6.8). The proximal domain is the hilar region where a Type 1 cell's primary cilium protrudes. Type 1 cells have additional proximal processes that extend laterally and contact blood vessels and other Type 1 cells. The intermediate domain includes the granule cell layer and SGZ. The Type 1 cell bodies are in close proximity with intermediate neural progenitors. The Type 1 cell primary dendrite is closely associated with mature granule cells as it extends through the granule cell layer. The distal domain consists of the inner molecular layer where the dendritic arbor of Type 1 cells branches. The dendrites receive GABAergic input from hilar interneurons and glutamatergic input from mossy cells. Physical contact and the secretion of signaling molecules between neurons, astrocytes, microglia, pericytes and endothelial cells enable niche interactions that regulate proliferation and differentiation of neurons (Cao et al., 2013; Ehret et al., 2015; Moss et al., 2016).

Small capillaries consist of endothelial cells whose tight connections form the blood-brain-barrier (BBB) that prevents substances in the blood from passing into the brain. Instead, transport molecules in endothelial cells serve to selectively allow necessary molecules to access the brain from the blood. Type 1 and neural progenitor cells are found in close proximity to endothelial cells. Type 1 cells use an endfoot process to wrap around the capillaries (Ehret et al., 2015; Stolp and Molnár, 2015). Endothelial cell secretions, such as brain-derived neurotrophic factor (BDNF) and vascular endothelial growth factor (VEGF), influence proliferation in the adult brain (Leventhal et al., 1999; Shen, 2004). NSCs divide symmetrically when they are cultured with endothelial cells, resulting in an expansion of the NSC population (Shen, 2004). When NSCs are cultured alone, they divide asymmetrically and produce neurons, astrocytes, and oligodendrocytes (Shen, 2004). Thus endothelial cells are an important regulator of Type 1 cell proliferation and, as such, support NSC maintenance.

Pericytes densely cover endothelial cells in the hippocampal niche (Ehret et al., 2015). Proliferation of NSCs is increased when they are cocultured with pericytes, though the effect on cell growth and differentiation is negligible (Ehret et al., 2015). Further, the majority of NSCs cultured with pericytes differentiate into glia. This little-studied population may mediate cohesion of niche cells, but more research is needed to determine the additional roles of pericytes (Ehret et al., 2015).

Microglia are the immune cells of the brain, existing in a surveillant state under normal conditions. Inflammation is a natural response in the brain to disease, injury, or infection, resulting in the release of proinflammatory cytokines from microglia such as interleukin 1 β, interleukin 6, and tumor necrosis factor α. Microglia protect healthy cells by phagocytosing necrotic or apoptotic cells. During adult neurogenesis, an average of two-thirds of the generated Type 3 cells become apoptotic naturally and are eliminated by microglia (see below; Sierra et al., 2014; Urbán and Guillemot, 2014).

Glial, non-stem cell astrocytes are another important cell type in the niche. Astrocytes are GFAP-positive, but negative for Nestin and have a different morphology than GFAP-positive, Type 1 NSCs (Fig. 6.9). Early studies showed that when NSCs isolated from the adult hippocampus were cultured with astrocytes, the stem cells adopted a neuronal fate (Song et al., 2002). Astrocytes surround NSCs and release the proliferative factor 5′-triphosphate (ATP), which activates the inositol 1,4,5-trisphosphate receptor type 2 (IP3 R2) on NSCs (Cao et al., 2013; Ehret et al., 2015). Astrocytes also control the maturation of dendritic length and spine density, which directly affects the survival of adult-born neurons (Sultan et al., 2015). Further, astrocytes release D-serine, an NMDA-receptor co-agonist that increases the NMDA receptor's excitatory postsynaptic current that also has a positive influence on survival and integration (Sultan et al., 2015).

The neurons themselves are an important component of the niche, and are often the source of secretory molecules affecting nearby cells. For example, NSCs and progenitor cells express CXCR4, while the mature granule cells express CXCL12 (see Chapter 3). CXCR4 expression is necessary for subsequent expression of DCX, placement of new neurons in the inner third of the granule cell layer and development of a normal dendritic arbor (Schultheiß et al., 2013).

Though quiescent, Type 1 NSCs have an important niche role associated with synaptic activity. The primary dendritic process of the Type 1 cell twists and turns, ensheathing local synapses and vasculature in the inner molecular layer as it extends from the SGZ through the granule cell layer (Moss et al., 2016). Type 1 cells may form an adherence with an adjacent astrocytic process and both processes then ensheath a single synapse. The physical integration of Type 1 cells in the niche enables bidirectional signaling and regulation of surrounding cells. The interaction also makes Type 1 cells sensitive to synaptic activity. Hyperactivity such as sustained epileptic seizures causes the NSCs to divide symmetrically, and both mother and daughter cell convert into reactive astrocytes, eventually depleting NSCs (Sierra et al., 2015).

REGULATING THE BEHAVIORS OF ADULT-BORN CELLS

Each phase of adult neurogenesis is regulated or influenced by several factors. In the following, we review the contemporary research on NSC maintenance, quiescent Type 1 cell activation and differentiation, proliferation of NSCs and intermediate progenitors, survival, maturation, and integration.

NSC Maintenance

Type 1 cells are typically quiescent, which protects them from metabolic stress and maintains the genomic integrity (Cheung and Rando, 2013). Quiescence is reversible and must be balanced with self-renewal to ensure a stable population of Type 1 cells (Simons and Clevers, 2011). Similar to hair follicle stem cells (Chapter 3), BMP signaling maintains Type 1 cells in a quiescent state (for review, see Fig. 6.1, 6.2, 6.4, and 6.6–6.8). Activation of BMP receptors by BMP ligands results in the phosphorylation and nuclear translocation of Smad1, and these cells remain quiescent (Mira et al., 2010). Similar to germline stem cells in *Caenorhabditis elegans*, Notch signaling results in Type 1 renewal (Ables et al., 2010). Hair follicle stem cells have an ectodermal origin, and germline stem cells arise from the endoderm; yet there is a conservation of signaling pathways for regulating quiescence and self-renewal in these diverse stem cell types.

Activation of Quiescent Cells

The proneural transcription factor *achaete-scute homolog 1* (Ascl1, also known as Mash1) has well-characterized roles in embryonic neurogenesis, including proliferation, specification and differentiation of neurons (Bertrand et al., 2002; Castro et al., 2011). Ascl1 is a significant regulator of Type 1 cell activation, expressed in both Type 1 and intermediate neural progenitors, and specifically, in cycling stem cells (Breunig et al., 2007; Kim et al., 2011). As a transcription factor, Ascl1 directly controls proliferation by activating cell cycle regulator genes. The number of Ascl1-expressing, activated Type 1 cells is increased by glutamate receptor activation or Notch signaling inhibition (Andersen et al., 2014). Another study found that a protein called Huwe1 (HECT, UBA and WWE domain containing 1) destabilizes Ascl1 and returns Type 1 cells to a quiescent state (Urbán et al., 2016). Thus activation of Type 1 cells is balanced by regulating the activity of Ascl1.

GABAergic signaling from local interneurons is another mechanism to return Type 1 cells to a quiescent state. There are multiple interneuron

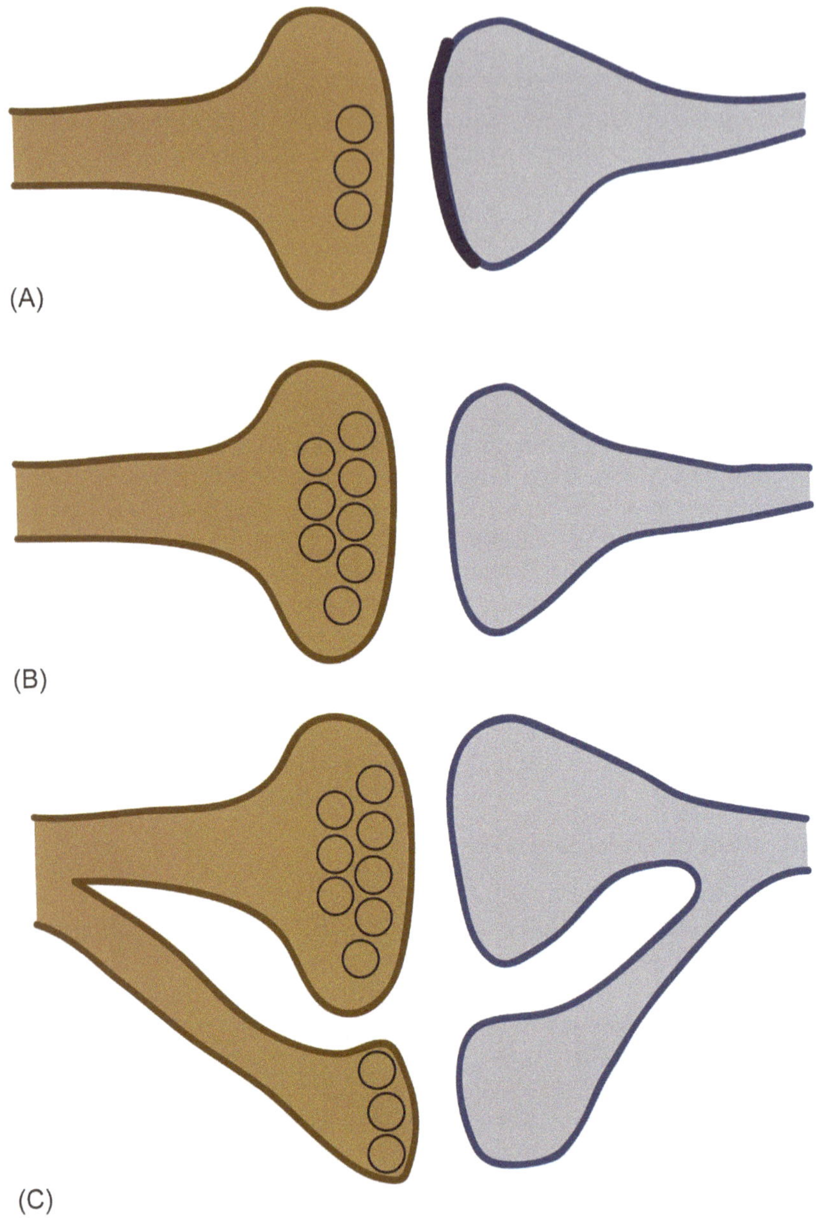

FIG. 6.7 LTP is a biological process of plasticity that strengthens the synapses between neurons. Plasticity may occur in three nonexclusive ways. (A) More neurotransmitter receptors are inserted in the postsynaptic membrane. (B) More neurotransmitter is released from the presynaptic terminal. (C) The synapse can undergo extensive remodeling to increase the synaptic area.

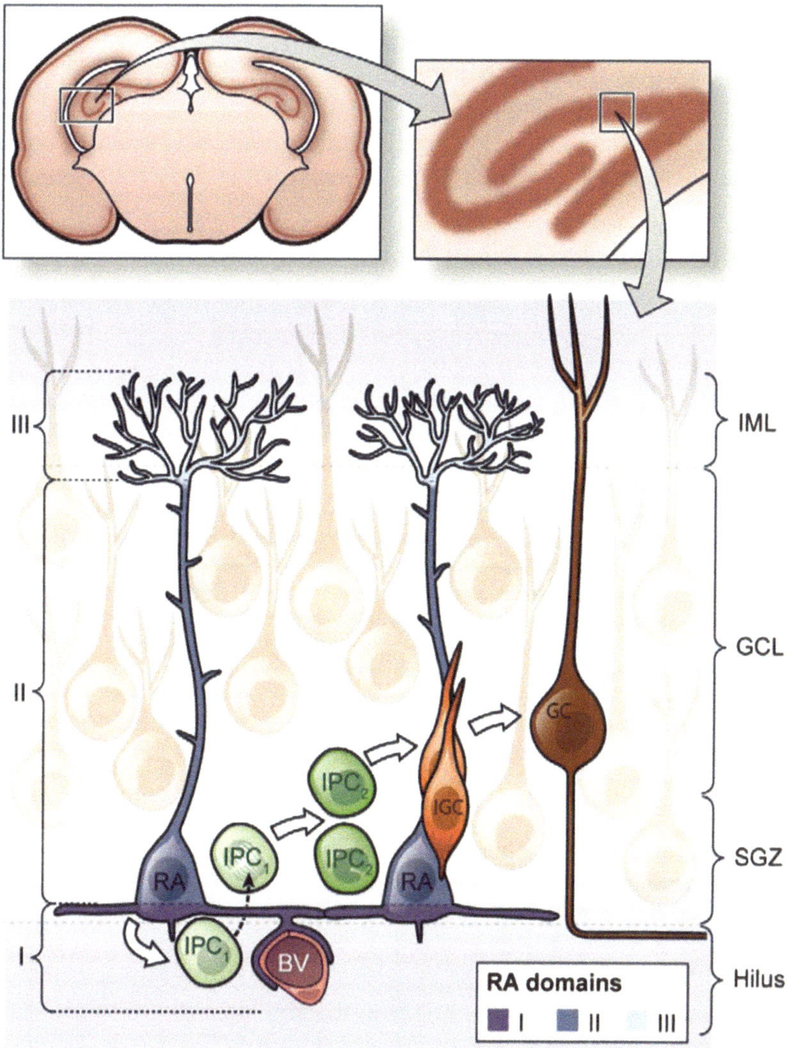

FIG. 6.8 Anatomy of the domains of Type 1 NSCs in the dentate gyrus of the hippocampus. (Top) The location of the hippocampus is highlighted in a coronal section of an adult mouse brain (left). The dentate gyrus looks like a backwards "c" in the hippocampal formation (right). (Bottom) Detailed cellular domains of the box from the Top right figure. In the Proximal (I) domain, the Type 1 cells each extend a primary cilium and blood vessel (BV)-associated endfeet. Lateral processes contact neighboring Type 1 cells. Intermediate progenitor cells (IPC) are also observed in the proximal domain. In the Intermediate (II) domain, Type 1 cells (radial astrocytes, RA) give rise to the IPCs, Type 2a and 2b (IPC_1, IPC_2). The Type 3 neuroblasts (immature granule cells, IGC) migrate using the primary dendrite of the Type 1 cells a short distance before differentiating into an immature granule cell (GC). In the Distal (III) domain, immature granule cells develop a dendritic arbor and integrate into the existing circuitry alongside mature granule cells. *RA*, radial astrocytes. *Reprinted from Fuentealba et al. (2012) with permission from Elsevier.*

populations in the adult dentate gyrus such as calretinin-, somatostatin-, cholecystokinin-, and parvalbumin (PV)-positive types. The PV-positive interneurons specifically release GABA onto Type 1 cells that induces the NSCs into quiescence. Glutamatergic activation and GABAergic inhibition demonstrate a neuron-specific niche interaction that regulates Type 1 cell behavior (Song et al., 2012b).

Proliferation of Intermediate Neural Progenitors

In the SGZ niche, VEGF is secreted by Type 1 and progenitor cells, and is known to promote neurogenesis and angiogenesis by stimulating the proliferation of vascular endothelial cells (Jin et al., 2002; Kirby et al., 2015; Rosenstein et al., 1998). The fact that VEGF is secreted by Type 1 cells suggests that they maintain their own niche by protein secretion (Kirby et al., 2015). Indeed, the loss of VEGF results in impaired stem cell maintenance in culture (Kirby et al., 2015). Intracranioventricular injections of VEGF doubled the number of BrdU-positive cells in Sox2-expressing Type 1 and Type 2a cells (Fournier et al., 2012). Thus VEGF is an important factor for promoting proliferation in NSCs (Fournier et al., 2012).

Wnt/β-catenin signaling was introduced in Chapter 3 as an important signal for regenerative (anagen) phase in hair follicles. In the dentate gyrus, Wnt3 signaling is sufficient to stimulate proliferation, specifically targeting the Type 3 neuroblast cells (Lie et al., 2005). In the absence of Wnt3 there was a significant decrease in the Type 3 cells, suggesting that Wnt3 may also be an important survival signal (Lie et al., 2005). Though enrichment is known to increase adult-born cell survival, Wn3 activity and enrichment have not been linked, though the timing suggests Wnt3 signaling could play a role (Tashiro et al., 2007).

Regulating Differentiation

Important progress has been made in the ability to identify differences between the proliferating cells of the hippocampus, Type 1, 2a, 2b, and 3 due to the development of immunohistochemical markers specific to each type that complements detailed morphological studies. A Type 1 cell has three fate choices that involve quiescence, differentiation or self-renewal (Bonaguidi et al., 2011). The first decision is whether to remain quiescent or become activated and enter the cell cycle. The second choice is how to divide when it is in an activated state. It can either (1) divide symmetrically, self-renew and expand the Type 1 cell pool or (2) undergo neurogenic asymmetric self-renewal that produces one new Type 1 cell and one new intermediate progenitor, or (3) under astrogliogenic asymmetric self-renewal to generate one astroglia cell and one Type 1 cell (Fig. 6.9). The third choice for an activated Type 1 cell is to either return to a quiescent

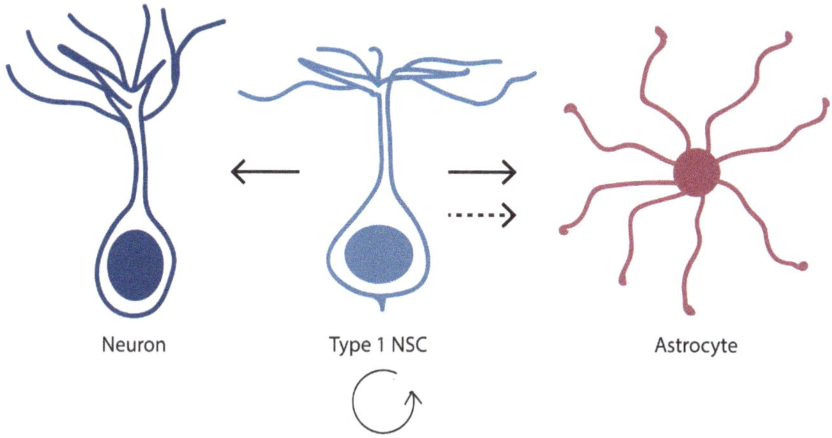

FIG. 6.9 The fates of a dividing Type 1 NSC. A Type 1 cell can divide symmetrically to expand the Type 1 cell pool. Alternatively, it can divide asymmetrically and produce either one new intermediate progenitor that will eventually differentiate into a neuron or one astrocyte (solid arrows). It is also possible for an activated or quiescent Type 1 cell to differentiate into an astrocyte (dashed arrow).

state or directly differentiate into an astrocyte without cell division. It is also possible for a *quiescent* Type 1 cell to differentiate into an astrocyte (Bonaguidi et al., 2011).

Epigenetics also plays an important role in the ability of an adult-born cell to differentiate through the five phases of adult neurogenesis before becoming an integrated, mature granule cell. Intermediate neural progenitors are known to proliferate for a time, but it is only recently that ideas regarding how that time is controlled have been developed. In vertebrates, methylation occurs only at cytosine residues adjacent to guanine residues with a phosphate group between them (CpG). The Methyl-CpG binding protein 1 (Mbd1) binds specifically to methylated DNA and promotes adult neurogenesis by binding the promoter of fibroblast growth factor 2 (Fgf2), an intermediate neural progenitor mitogen (Li et al., 2008; Ma et al., 2010). Fgf2 expression promotes proliferation in intermediate neural progenitors, but only for a time. While Mbd1 "reads" CpG-methylated DNA, emerging evidence suggests that it also affects DNA methylation levels. One theory is that Mbd1 recruits proteins that further methylate Fgf2 DNA in a manner that silences the gene over many rounds of cell division that determines the time that a cell can proliferate. Eventually the intermediate progenitor will no longer proliferate and proceed to the next phase of differentiation. Thus Mbd1 may create a form of "cellular memory" (Ma et al., 2010).

Survival

During development as many as 50% of neurons produced normally die by caspase 3-dependent apoptosis (Morshead and van der Kooy, 1992; Pompeiano et al., 2000). In the dentate gyrus an average of two-thirds of intermediate neural progenitors undergo apoptosis, a major regulatory checkpoint of adult neurogenesis (Urbán and Guillemot, 2014). There is likely a competition for specific resources such as synaptic connections that allow integration into the existing circuitry that may provide survival signals. Understanding adult-born neuron survival has important implications for the development of cell-replacement therapy following injury or disease (Bergami and Berninger, 2012).

Microglia exist in a surveillant state unless activated due to inflammation that may occur naturally due to aging or Alzheimer disease, or artificially activated by irradiation or treatment with the bacterial lipopolysaccharides (LPS) (Sierra et al., 2014). During inflammation, microglia, as well as astrocytes, release interleukin-6, a proinflammatory cytokine that causes apoptosis of the Type 3 neuroblasts. Surveillant microglia phagocytose the apoptotic cells (Sierra et al., 2010). Treatment with VEGF, known to increase proliferation, does not reduce the number of apoptotic cells (Jin et al., 2002). Instead the number of apoptotic cells increases, though not as greatly as the number of proliferating cells (Jin et al., 2002). These data suggest that proliferation and survival each have distinct regulatory checkpoints.

The survival phase of adult neurogenesis is sensitive to hippocampus-dependent activity such as associative learning (classical conditioning) or exposure to enriched environments (spatial complexity) (Gould et al., 1999). In rats that underwent associative or spatial learning, there was a significant increase in the number of BrdU-positive cells that also expressed Turned-on-after-division (TOAD), a marker of mature neurons (Gould et al., 1999). Sixty days after associative learning had stopped, the neurons were still present (Leuner, 2004). These data suggest that engaging in hippocampus-dependent activities increases the survival, integration and maturation of adult-born neurons.

Rho-GTPases play an important role in cytoskeletal rearrangement that is necessary during axon and dendrite development referred to as neurite outgrowth. Formation of dendritic synapses is a necessary requirement for adult-born neurons to survive (Bergami and Berninger, 2012). Rho-GTPases are inactive when bound by GDP, and require a guanine nucleotide exchange factor (GEF) to replace GDP with GTP. Once Rho-GTPase is GTP-bound, it is active and can contribute to cytoskeletal rearrangement. The GTPase activating protein (GAP) returns Rho-GTPases to the inactive state by hydrolyzing GTP, resulting in the removal of a phosphate group, and the conversion of GTP to GDP (Fig. 6.10).

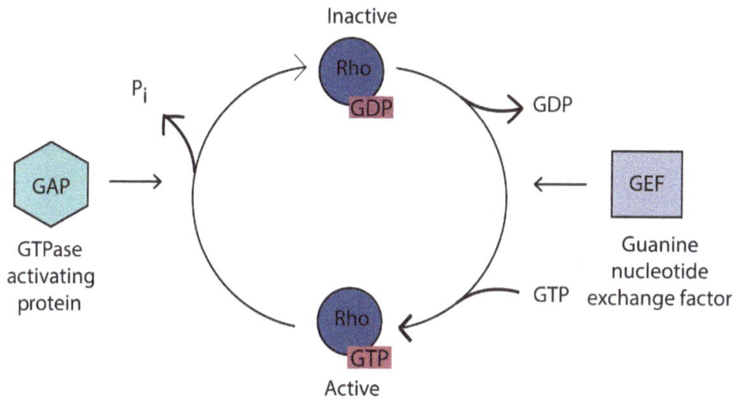

FIG. 6.10 Rho GTPase regulation by GTPase activating protein (GAP) and guanine nucleotide exchange factor (GEF). When Rho GTPase is bound to GDP (top) it is in an inactive state. GEF facilitates the transfer of GDP for GTP (right), which converts Rho GTPase to its active form (bottom). GAP hydrolyzes GTP (left) resulting in the loss of a phosphate group and converts Rho GTPase to its inactive form.

There are specific Rho-GTPases for neurite growth (Rac1) and neurite retraction (RhoA). Both extension and retraction are necessary as a neurite develops into a permanent part of the neuron. RhoA facilitates retraction through the Rho-associated kinase (ROCK). By inhibiting RhoA and thus, neurite retraction, one study showed an increase in adult-born neuron survival, emphasizing the importance of neurite development for survival (Christie et al., 2013).

Maturation and Integration

Once cells exit the cell cycle they extend an axon to CA3 that establishes contact within 10 days of cell division (Gould et al., 1999). Concurrently, dendritic processes begin to develop in the granule cell layer (Zhao, 2006). After 2 weeks, the new cells resemble a mature granule cell neuron with a dendritic arbor in the inner molecular layer. Over the next week, the dendritic arbor expands into the middle and outer molecular layer. From 21 to 28 days after birth, dendritic spines develop, which is where the majority of glutamatergic input is received (Zhao, 2006). An average granule cell has a dendritic arbor totaling 3000 μm in length with an average spine density of 2/μm for a total of 6000 spines/cell (Schmidt-Hieber et al., 2004).

GABA persists in having a depolarizing effect on the cells due to the expression of NKCC1 and a high internal Cl^- gradient for the first 2 weeks (Ge et al., 2006). Depolarization results in Ca^{2+} influx that is necessary for the expression of NeuroD, a marker of neuronal fate differentiation

(Tozuka et al., 2005). After 2 weeks the levels of NKCC1 decrease and the K^+-Cl^--cotransporter KCC2 expression is increased. The change in cotransporters alters the internal Cl^- concentration resulting in GABA now having an inhibitory (hyperpolarizing) effect (Leonzino et al., 2016).

Early maturing neurons begin to express voltage-gated channels within 2–3 weeks of birth as well as the T-Type Ca^{2+} channels (Couillard-Despres et al., 2006; Schmidt-Hieber et al., 2004). The T-type channels have a low threshold of activation, and because of the high density of voltage-gated sodium channels present, are able to elicit action potentials with only a small amount of excitation (Couillard-Despres et al., 2006). Thus immature neurons are more susceptible to LTP induction (Box 6.2).

The first glutamatergic synapses coincide with the expression of voltage-gated channel expression (2–3 weeks), forming on the developing dendritic spines (Ge et al., 2006; Zhao, 2006). The new neurons begin to express NMDA receptors that are important for neuron survival and integration (Tashiro et al., 2007). NMDA receptors are an essential part of LTP induction (Box 6.2), and contribute to the unique properties of immature neurons (Tashiro et al., 2006).

During the early maturation stage the new neuron is thought to play an important role in the processes of learning and memory. There are four- to fivefold more neurons in the dentate gyrus than in either the entorhinal cortex or CA3. Only a small subset of granule cells are activated by any given event (Chawla et al., 2005; Jung and McNaughton, 1993). In this way, the dentate gyrus neurons act to separate patterns of activation—partially decorrelating the input before sending the signal on to CA3 (Marr, 1971; O'Reilly and McClelland, 1994). Separating patterns of input allows similar memories to be differentiated (Box 6.2). The enhanced plasticity and greater excitability of immature neurons is balanced by their connections to inhibitory interneurons. Excitation of immature neurons causes the excitation of synaptically connected, inhibitory neurons that decreases activity in other neurons. The reduction in overall network activity provides sparse representation of a memory that reduces the interference between other memory representations (Ikrar et al., 2013). Indeed, increasing the survival of adult-born neurons was directly correlated with improved pattern separation (cognitive tasks). Thus adult-born neurons are important for learning and memory.

SUMMARY

Adult neurogenesis in the hippocampus occurs throughout life. NSCs give rise to neurons and glia, though only a percentage of those produced survive. Over 6–8 weeks it takes for an adult-born cell to become a fully functional neuron, the expression of specific genes is highly orchestrated. Each gene encodes a protein that plays a specific role in the process of adult neurogenesis.

Surrounding the NSCs, progenitors, immature and mature neurons are cells of the niche that contribute to the support of and/or regulate aspects of adult neurogenesis. Interactions between the adult-born cells and other niche cells determine the levels of proliferation, survival and integration.

The connections between the hippocampus and other structures in the limbic system are extensive. Not surprisingly, the function of the hippocampus and adult neurogenesis is affected by mental states such as anxiety and depression. While aging has been extensively documented to decrease adult neurogenesis, exercise, learning, environmental enrichment and a number of antidepressant drugs have all been shown to increase neurogenesis. Further research is required on how states of health and disease affect the regulation of adult neurogenesis, either by affecting the adult-born cells directly, or through a population of niche cells. In Chapter 8, we will discuss the evidence and correlations between states of health and disease, aging and adult neurogenesis.

DEFINITIONS

Achaete-scute homolog 1 (Ascl1) A proneural gene that controls the specification of neural progenitors and later steps of differentiation and neurite outgrowth.

α-Amino-3-hydroxy-5-methyl-4-isoxazolepropionic acid (AMPA) An amino acid whose structure is similar to glutamate. AMPA binds to a receptor that activates glutamate-gated ion channels, referred to as AMPA receptors.

Brain lipid binding protein (BLBP) Contributes to fatty acid uptake, transport and metabolism.

Calbindin An intracellular protein and member of the EF-hand proteins (transcription factor category) with high affinity for calcium. Contributes to calcium buffering and synaptic plasticity.

Calretinin A calcium binding protein and member of the EF-hand proteins. Contributes to calcium buffering and modulates intrinsic neuronal excitability.

Dendritic spines A small protrusion from a dendrite that contains all the elements of a postsynaptic terminal. Each spine typically forms a synapse with a single axon.

Doublecortin (DCX) A microtubule-associated protein that stabilizes and causes bundling of microtubules. A common marker of migratory and immature neurons.

Engram	Persistent changes in synaptic strength via LTP or LTD of preexisting connections as a mechanism for memory formation.
Gamma-aminobutyric acid (GABA)	Neurons that express the neurotransmitter GABA are called GABAergic. There are two types of GABA receptors, $GABA_A$ and $GABA_B$. $GABA_A$ receptors gate ionotropic Cl^- channels and are sensitive to bicuculline antagonism. $GABA_B$ receptors are G protein-coupled receptors that conduct an inward K+ current and can be blocked by baclofen.
Glial fibrillary acidic protein (GFAP)	A member of the intermediate filament family expressed in astroglial cells. Most likely involved in controlling the shape, movement, and function of astrocytes.
***N*-Methyl-D-aspartate (NMDA)**	An amino acid whose structure is similar to glutamate. NMDA binds to a receptor that activates glutamate-gated ion channels, referred to as NMDA receptors.
Nestin	An intermediate filament protein expressed in dividing cells; a marker of self-renewal. Expressed in the majority of mitotically active CNS progenitors that give rise to neurons and glia.
NeuN	A neuronal nuclear antigen commonly used as a biomarker for neurons. A homolog to sex-determining genes in *C. elegans*.
NeuroD	A proneural basic helix-loop-helix (bHLH) transcription factor that induces neuronal differentiation.
Prospero-related homeobox 1 (Prox-1)	Promotes self-renewal of embryonic NSCs, neurogenesis, and inhibits astrogliogenesis. Suppresses cell cycle progression via a direct action in basic components of the cell cycle machinery.
Polysialic neural cell adhesion molecule (PSA-NCAM)	The polysialylated form of the cell surface glycoprotein, neural cell adhesion molecule. A marker of developing and migratory neurons.
Sox2	A transcription factor that maintains pluripotency and inhibits neuronal differentiation.
T-box brain protein 2 (Tbr2)	A transcription factor also known as Eomesodermin. Tbr2 is required for establishing neuronal progenitors in the adult SGZ.
Tuj1	A neuron-specific Class III β-tubulin that is present in intermediate neural progenitors and immature granule cells.

References

Ables, J.L., DeCarolis, N.A., Johnson, M.A., Rivera, P.D., Gao, Z., Cooper, D.C., Radtke, F., Hsieh, J., Eisch, A.J., 2010. Notch1 is required for maintenance of the reservoir of adult hippocampal stem cells. J. Neurosci. 30, 10484–10492.

Andersen, J., Urbán, N., Achimastou, A., Ito, A., Simic, M., Ullom, K., Martynoga, B., Lebel, M., Göritz, C., Frisén, J., et al., 2014. A transcriptional mechanism integrating inputs from extracellular signals to activate hippocampal stem cells. Neuron 83, 1085–1097.

Arnold, S.J., Huang, G.-J., Cheung, A.F.P., Era, T., Nishikawa, S.-I., Bikoff, E.K., Molnár, Z., Robertson, E.J., Groszer, M., 2008. The T-box transcription factor Eomes/Tbr2 regulates neurogenesis in the cortical subventricular zone. Genes Dev. 22, 2479–2484.

Bergami, M., Berninger, B., 2012. A fight for survival: The challenges faced by a newborn neuron integrating in the adult hippocampus. Dev. Neurobiol. 72, 1016–1031.

Bertrand, N., Castro, D.S., Guillemot, F., 2002. Proneural genes and the specification of neural cell types. Nat. Rev. Neurosci. 3, 517–530.

Björklund, A., Stenevi, U., Schmidt, R.H., Dunnett, S.B., Gage, F.H., 1983. Intracerebral grafting of neuronal cell suspensions. I. Introduction and general methods of preparation. Acta Physiol. Scand. Suppl. 522, 1–7.

Bonaguidi, M.A., Wheeler, M.A., Shapiro, J.S., Stadel, R.P., Sun, G.J., Ming, G., Song, H., 2011. In vivo clonal analysis reveals self-renewing and multipotent adult neural stem cell characteristics. Cell 145, 1142–1155.

Bourne, J., Harris, K.M., 2007. Do thin spines learn to be mushroom spines that remember? Curr. Opin. Neurobiol. 17, 381–386.

Breunig, J.J., Silbereis, J., Vaccarino, F.M., Sestan, N., Rakic, P., 2007. Notch regulates cell fate and dendrite morphology of newborn neurons in the postnatal dentate gyrus. Proc. Natl. Acad. Sci. 104, 20558–20563.

Brundin, P., Barbin, G., Isacson, O., Mallat, M., Chamak, B., Prochiantz, A., Gage, F.H., Björklund, A., 1985. Survival of intracerebrally grafted rat dopamine neurons previously cultured in vitro. Neurosci. Lett. 61, 79–84.

Cabrera-Pastor, A., Llansola, M., Felipo, V., 2016. Extracellular protein kinase A modulates intracellular calcium/calmodulin-dependent protein kinase II, nitric oxide synthase, and the glutamate-nitric oxide-cGMP pathway in cerebellum. Differential effects in hyperammonemia. ACS Chem. Neurosci. 7, 1753–1759.

Cao, X., Li, L.-P., Qin, X.-H., Li, S.-J., Zhang, M., Wang, Q., Hu, H.-H., Fang, Y.-Y., Gao, Y.-B., Li, X.-W., et al., 2013. Astrocytic adenosine 5′-triphosphate release regulates the proliferation of neural stem cells in the adult hippocampus: mitogenic effect of astrocytic ATP in adult SGZ. Stem Cells 31, 1633–1643.

Castro, D.S., Martynoga, B., Parras, C., Ramesh, V., Pacary, E., Johnston, C., Drechsel, D., Lebel-Potter, M., Garcia, L.G., Hunt, C., et al., 2011. A novel function of the proneural factor Ascl1 in progenitor proliferation identified by genome-wide characterization of its targets. Genes Dev. 25, 930–945.

Chawla, M.K., Guzowski, J.F., Ramirez-Amaya, V., Lipa, P., Hoffman, K.L., Marriott, L.K., Worley, P.F., McNaughton, B.L., Barnes, C.A., 2005. Sparse, environmentally selective expression of Arc RNA in the upper blade of the rodent fascia dentata by brief spatial experience. Hippocampus 15, 579–586.

Cheung, T.H., Rando, T.A., 2013. Molecular regulation of stem cell quiescence. Nat. Rev. Mol. Cell Biol. 14, 329–340.

Christie, K.J., Turbic, A., Turnley, A.M., 2013. Adult hippocampal neurogenesis, Rho kinase inhibition and enhancement of neuronal survival. Neuroscience 247, 75–83.

Cohen, N., Squire, L., 1980. Preserved learning and retention of pattern-analyzing skill in amnesia: dissociation of knowing how and knowing that. Science 210, 207–210.

Couillard-Despres, S., Winner, B., Karl, C., Lindemann, G., Schmid, P., Aigner, R., Laemke, J., Bogdahn, U., Winkler, J., Bischofberger, J., et al., 2006. Targeted transgene expression in neuronal precursors: watching young neurons in the old brain. Eur. J. Neurosci. 24, 1535–1545.

Deng, W., Aimone, J.B., Gage, F.H., 2010. New neurons and new memories: how does adult hippocampal neurogenesis affect learning and memory? Nat. Rev. Neurosci. 11, 339–350.

Doucet, M.V., Harkin, A., Dev, K.K., 2012. The PSD-95/nNOS complex: new drugs for depression? Pharmacol. Ther. 133, 218–229.

Dranovsky, A., Leonardo, E.D., 2012. Is there a role for young hippocampal neurons in adaptation to stress? Behav. Brain Res. 227, 371–375.

Duan, X., Chang, J.H., Ge, S., Faulkner, R.L., Kim, J.Y., Kitabatake, Y., Liu, X., Yang, C.-H., Jordan, J.D., Ma, D.K., et al., 2007. Disrupted-in-schizophrenia 1 regulates integration of newly generated neurons in the adult brain. Cell 130, 1146–1158.

Ehret, F., Vogler, S., Kempermann, G., 2015. A co-culture model of the hippocampal neurogenic niche reveals differential effects of astrocytes, endothelial cells and pericytes on proliferation and differentiation of adult murine precursor cells. Stem Cell Res. 15, 514–521.

Eichenbaum, H., 2014. Time cells in the hippocampus: a new dimension for mapping memories. Nat. Rev. Neurosci. 15, 732–744.

Encinas, J.M., Vaahtokari, A., Enikolopov, G., 2006. Fluoxetine targets early progenitor cells in the adult brain. Proc. Natl. Acad. Sci. 103, 8233–8238.

Fournier, N.M., Lee, B., Banasr, M., Elsayed, M., Duman, R.S., 2012. Vascular endothelial growth factor regulates adult hippocampal cell proliferation through MEK/ERK- and PI3K/Akt-dependent signaling. Neuropharmacology 63, 642–652.

Fukuda, S., Kato, F., Tozuka, Y., Yamaguchi, M., Miyamoto, Y., Hisatsune, T., 2003. Two distinct subpopulations of nestin-positive cells in adult mouse dentate gyrus. J. Neurosci. 23, 9357–9366.

Ge, S., Goh, E.L.K., Sailor, K.A., Kitabatake, Y., Ming, G., Song, H., 2006. GABA regulates synaptic integration of newly generated neurons in the adult brain. Nature 439, 589–593.

Ge, S., Yang, C.-H., Hsu, K.-S., Ming, G.-L., Song, H., 2007. A critical period for enhanced synaptic plasticity in newly generated neurons of the adult brain. Neuron 54, 559–566.

Gould, E., Beylin, A., Tanapat, P., Reeves, A., Shors, T.J., 1999. Learning enhances adult neurogenesis in the hippocampal formation. Nat. Neurosci. 2, 260–265.

Hardingham, N., Dachtler, J., Fox, K., 2013. The role of nitric oxide in pre-synaptic plasticity and homeostasis. Front. Cell. Neurosci. 7.

Hastings, N.B., Gould, E., 1999. Rapid extension of axons into the CA3 region by adult-generated granule cells. J. Comp. Neurol. 413, 146–154.

Hebb, D., 1949. The Organization of Behavior. A Neuropsychological Theory. Wiley, New York, NY.

Hodge, R.D., Kowalczyk, T.D., Wolf, S.A., Encinas, J.M., Rippey, C., Enikolopov, G., Kempermann, G., Hevner, R.F., 2008. Intermediate progenitors in adult hippocampal neurogenesis: Tbr2 expression and coordinate regulation of neuronal output. J. Neurosci. 28, 3707–3717.

Hodge, R.D., Nelson, B.R., Kahoud, R.J., Yang, R., Mussar, K.E., Reiner, S.L., Hevner, R.F., 2012. Tbr2 is essential for hippocampal lineage progression from neural stem cells to intermediate progenitors and neurons. J. Neurosci. 32, 6275–6287.

Huttmann, K., Sadgrove, M., Wallraff, A., Hinterkeuser, S., Kirchhoff, F., Steinhauser, C., Gray, W.P., 2003. Seizures preferentially stimulate proliferation of radial glia-like astrocytes in the adult dentate gyrus: functional and immunocytochemical analysis. Eur. J. Neurosci. 18, 2769–2778.

Ikrar, T., Guo, N., He, K., Besnard, A., Levinson, S., Hill, A., Lee, H.-K., Hen, R., Xu, X., Sahay, A., 2013. Adult neurogenesis modifies excitability of the dentate gyrus. Front. Neural Circ. 7.

Jin, K., Zhu, Y., Sun, Y., Mao, X.O., Xie, L., Greenberg, D.A., 2002. Vascular endothelial growth factor (VEGF) stimulates neurogenesis in vitro and in vivo. Proc. Natl. Acad. Sci. 99, 11946–11950.

Jonides, J., Lewis, R.L., Nee, D.E., Lustig, C.A., Berman, M.G., Moore, K.S., 2008. The mind and brain of short-term memory. Annu. Rev. Psychol. 59, 193–224.

Jung, M.W., McNaughton, B.L., 1993. Spatial selectivity of unit activity in the hippocampal granular layer. Hippocampus 3, 165–182.

Karalay, O., Doberauer, K., Vadodaria, K.C., Knobloch, M., Berti, L., Miquelajauregui, A., Schwark, M., Jagasia, R., Taketo, M.M., Tarabykin, V., et al., 2011. Prospero-related homeobox 1 gene (Prox1) is regulated by canonical Wnt signaling and has a stage-specific role in adult hippocampal neurogenesis. Proc. Natl. Acad. Sci. 108, 5807–5812.

Kempermann, G., 2003. Early determination and long-term persistence of adult-generated new neurons in the hippocampus of mice. Development 130, 391–399.

Kempermann, G., 2015. Astrocytes, makers of new neurons. Neuron 88, 850–851.

Kempermann, G., Jessberger, S., Steiner, B., Kronenberg, G., 2004. Milestones of neuronal development in the adult hippocampus. Trends Neurosci. 27, 447–452.

Kim, E.J., Ables, J.L., Dickel, L.K., Eisch, A.J., Johnson, J.E., 2011. Ascl1 (Mash1) defines cells with long-term neurogenic potential in subgranular and subventricular zones in adult mouse brain. PLoS ONE 6, e18472.

Kirby, E.D., Kuwahara, A.A., Messer, R.L., Wyss-Coray, T., 2015. Adult hippocampal neural stem and progenitor cells regulate the neurogenic niche by secreting VEGF. Proc. Natl. Acad. Sci. U. S. A. 112, 4128–4133.

Komorowski, R.W., Manns, J.R., Eichenbaum, H., 2009. Robust conjunctive item-place coding by hippocampal neurons parallels learning what happens where. J. Neurosci. 29, 9918–9929.

Kronenberg, G., Reuter, K., Steiner, B., Brandt, M.D., Jessberger, S., Yamaguchi, M., Kempermann, G., 2003. Subpopulations of proliferating cells of the adult hippocampus respond differently to physiologic neurogenic stimuli. J. Comp. Neurol. 467, 455–463.

Leavitt, B.R., Hernit-Grant, C.S., Macklis, J.D., 1999. Mature astrocytes transform into transitional radial glia within adult mouse neocortex that supports directed migration of transplanted immature neurons. Exp. Neurol. 157, 43–57.

Leonzino, M., Busnelli, M., Antonucci, F., Verderio, C., Mazzanti, M., Chini, B., 2016. The timing of the excitatory-to-inhibitory GABA switch is regulated by the oxytocin receptor via KCC2. Cell Rep. 15, 96–103.

Leuner, B., 2004. Learning enhances the survival of new neurons beyond the time when the hippocampus is required for memory. J. Neurosci. 24, 7477–7481.

Leutgeb, J.K., Leutgeb, S., Moser, M.-B., Moser, E.I., 2007. Pattern separation in the dentate gyrus and CA3 of the hippocampus. Science 315, 961–966.

Leventhal, C., Rafii, S., Rafii, D., Shahar, A., Goldman, S.A., 1999. Endothelial trophic support of neuronal production and recruitment from the adult mammalian subependyma. Mol. Cell. Neurosci. 13, 450–464.

Li, X., Barkho, B.Z., Luo, Y., Smrt, R.D., Santistevan, N.J., Liu, C., Kuwabara, T., Gage, F.H., Zhao, X., 2008. Epigenetic regulation of the stem cell mitogen Fgf-2 by Mbd1 in adult neural stem/progenitor cells. J. Biol. Chem. 283, 27644–27652.

Lie, D.-C., Colamarino, S.A., Song, H.-J., Désiré, L., Mira, H., Consiglio, A., Lein, E.S., Jessberger, S., Lansford, H., Dearie, A.R., et al., 2005. Wnt signalling regulates adult hippocampal neurogenesis. Nature 437, 1370–1375.

Lucassen, P.J., Meerlo, P., Naylor, A.S., van Dam, A.M., Dayer, A.G., Fuchs, E., Oomen, C.A., Czéh, B., 2010. Regulation of adult neurogenesis by stress, sleep disruption, exercise and inflammation: implications for depression and antidepressant action☆. Eur. Neuropsychopharmacol. 20, 1–17.

Lynch, G., Deadwyler, S., Cotman, C., 1973. Postlesion axonal growth produces permanent functional connections. Science 180, 1364–1366.

Lynch, G., Larson, J., Kelso, S., Barrionuevo, G., Schottler, F., 1983. Intracellular injections of EGTA block induction of hippocampal long-term potentiation. Nature 305, 719–721.

Ma, D.K., Marchetto, M.C., Guo, J.U., Ming, G., Gage, F.H., Song, H., 2010. Epigenetic choreographers of neurogenesis in the adult mammalian brain. Nat. Neurosci. 13, 1338–1344.

Manns, J.R., Hopkins, R.O., Squire, L.R., 2003. Semantic memory and the human hippocampus. Neuron 38, 127–133.

Marr, D., 1971. Simple memory: a theory for archicortex. Philos. Trans. R. Soc. Lond. Ser. B Biol. Sci. 262, 23–81.

Maviel, T., 2004. Sites of neocortical reorganization critical for remote spatial memory. Science 305, 96–99.

Mira, H., Andreu, Z., Suh, H., Lie, D.C., Jessberger, S., Consiglio, A., San Emeterio, J., Hortigüela, R., Marqués-Torrejón, M.Á., Nakashima, K., et al., 2010. Signaling through BMPR-IA regulates quiescence and long-term activity of neural stem cells in the adult hippocampus. Cell Stem Cell 7, 78–89.

Morshead, C., van der Kooy, D., 1992. Postmitotic death is the fate of constitutively proliferating cells in the subependymal layer of the adult mouse brain. J. Neurosci. 12, 249–256.

Moser, M.-B., Rowland, D.C., Moser, E.I., 2015. Place cells, grid cells, and memory. Cold Spring Harb. Perspect. Biol. 7, a021808.

Moss, J., Gebara, E., Bushong, E.A., Sánchez-Pascual, I., O'Laoi, R., El M'Ghari, I., Kocher-Braissant, J., Ellisman, M.H., Toni, N., 2016. Fine processes of Nestin-GFP-positive radial glia-like stem cells in the adult dentate gyrus ensheathe local synapses and vasculature. Proc. Natl. Acad. Sci. 113, E2536–E2545.

Nagerl, U.V., Kostinger, G., Anderson, J.C., Martin, K.A.C., Bonhoeffer, T., 2007. Protracted synaptogenesis after activity-dependent spinogenesis in hippocampal neurons. J. Neurosci. 27, 8149–8156.

Nicola, Z., Fabel, K., Kempermann, G., 2015. Development of the adult neurogenic niche in the hippocampus of mice. Front. Neuroanat. 9.

O'Reilly, R.C., McClelland, J.L., 1994. Hippocampal conjunctive encoding, storage, and recall: avoiding a trade-off. Hippocampus 4, 661–682.

Owen, B., Grover, L.M., 2015. Activity-dependent differences in function between proximal and distal Schaffer collaterals. J. Neurophysiol. 113, 3646–3662.

Pompeiano, M., Blaschke, A.J., Flavell, R.A., Srinivasan, A., Chun, J., 2000. Decreased apoptosis in proliferative and postmitotic regions of the caspase 3-deficient embryonic central nervous system. J. Comp. Neurol. 423, 1–12.

Poo, M., Pignatelli, M., Ryan, T.J., Tonegawa, S., Bonhoeffer, T., Martin, K.C., Rudenko, A., Tsai, L.-H., Tsien, R.W., Fishell, G., et al., 2016. What is memory? The present state of the engram. BMC Biol. 14.

Raslau, F.D., Mark, L.P., Sabsevitz, D.S., Ulmer, J.L., 2015. Imaging of functional and dysfunctional episodic memory. Semin. Ultrasound CT MRI 36, 260–274.

Rickmann, M., Amaral, D.G., Cowan, W.M., 1987. Organization of radial glial cells during the development of the rat dentate gyrus. J. Comp. Neurol. 264, 449–479.

Rosenstein, J.M., Mani, N., Silverman, W.F., Krum, J.M., 1998. Patterns of brain angiogenesis after vascular endothelial growth factor administration in vitro and in vivo. Proc. Natl. Acad. Sci. U. S. A. 95, 7086–7091.

Scharfman, H.E., Myers, C.E., 2013. Hilar mossy cells of the dentate gyrus: a historical perspective. Front. Neural Circ. 6.

Schmidt-Hieber, C., Jonas, P., Bischofberger, J., 2004. Enhanced synaptic plasticity in newly generated granule cells of the adult hippocampus. Nature 429, 184–187.

Schultheiß, C., Abe, P., Hoffmann, F., Mueller, W., Kreuder, A.-E., Schütz, D., Haege, S., Redecker, C., Keiner, S., Kannan, S., et al., 2013. CXCR4 prevents dispersion of granule neuron precursors in the adult dentate gyrus: aberrant neurogenesis after Cxcr4 deletion. Hippocampus 23, 1345–1358.

Semon, R., 1921. The mneme. G. Allen & Unwin Ltd., The Macmillan Company, London, New York.

Seri, B., García-Verdugo, J.M., McEwen, B.S., Alvarez-Buylla, A., 2001. Astrocytes give rise to new neurons in the adult mammalian hippocampus. J. Neurosci. 21, 7153–7160.
Shapiro, M.L., Tanila, H., Eichenbaum, H., 1997. Cues that hippocampal place cells encode: dynamic and hierarchical representation of local and distal stimuli. Hippocampus 7, 624–642.
Shen, Q., 2004. Endothelial cells stimulate self-renewal and expand neurogenesis of neural stem cells. Science 304, 1338–1340.
Shi, S., 1999. Rapid spine delivery and redistribution of AMPA receptors after synaptic NMDA receptor activation. Science 284, 1811–1816.
Sierra, A., Encinas, J.M., Deudero, J.J.P., Chancey, J.H., Enikolopov, G., Overstreet-Wadiche, L.S., Tsirka, S.E., Maletic-Savatic, M., 2010. Microglia shape adult hippocampal neurogenesis through apoptosis-coupled phagocytosis. Cell Stem Cell 7, 483–495.
Sierra, A., Beccari, S., Diaz-Aparicio, I., Encinas, J.M., Comeau, S., Tremblay, M.-È., 2014. Surveillance, phagocytosis, and inflammation: how never-resting microglia influence adult hippocampal neurogenesis. Neural Plast. 2014, 1–15.
Sierra, A., Martín-Suárez, S., Valcárcel-Martín, R., Pascual-Brazo, J., Aelvoet, S.-A., Abiega, O., Deudero, J.J., Brewster, A.L., Bernales, I., Anderson, A.E., et al., 2015. Neuronal hyperactivity accelerates depletion of neural stem cells and impairs hippocampal neurogenesis. Cell Stem Cell 16, 488–503.
Silva, A.J., Paylor, R., Wehner, J.M., Tonegawa, S., 1992. Impaired spatial learning in alpha-calcium-calmodulin kinase II mutant mice. Science 257, 206–211.
Simons, B.D., Clevers, H., 2011. Strategies for homeostatic stem cell self-renewal in adult tissues. Cell 145, 851–862.
Smith, C.N., Squire, L.R., 2009. Medial temporal lobe activity during retrieval of semantic memory is related to the age of the memory. J. Neurosci. 29, 930–938.
Solstad, T., Boccara, C.N., Kropff, E., Moser, M.-B., Moser, E.I., 2008. Representation of geometric borders in the entorhinal cortex. Science 322, 1865–1868.
Song, H., Stevens, C.F., Gage, F.H., 2002. Astroglia induce neurogenesis from adult neural stem cells. Nature 417, 39–44.
Song, J., Christian, K.M., Ming, G., Song, H., 2012a. Modification of hippocampal circuitry by adult neurogenesis. Dev. Neurobiol. 72, 1032–1043.
Song, J., Zhong, C., Bonaguidi, M.A., Sun, G.J., Hsu, D., Gu, Y., Meletis, K., Huang, Z.J., Ge, S., Enikolopov, G., et al., 2012b. Neuronal circuitry mechanism regulating adult quiescent neural stem-cell fate decision. Nature 489, 150–154.
Steiner, B., Klempin, F., Wang, L., Kott, M., Kettenmann, H., Kempermann, G., 2006. Type-2 cells as link between glial and neuronal lineage in adult hippocampal neurogenesis. Glia 54, 805–814.
Stolp, H.B., Molnár, Z., 2015. Neurogenic niches in the brain: help and hindrance of the barrier systems. Front. Neurosci. 9.
Suh, H., Consiglio, A., Ray, J., Sawai, T., D'Amour, K.A., Gage, F.H., 2007. In vivo fate analysis reveals the multipotent and self-renewal capacities of Sox2+ neural stem cells in the adult hippocampus. Cell Stem Cell 1, 515–528.
Sultan, S., Li, L., Moss, J., Petrelli, F., Cassé, F., Gebara, E., Lopatar, J., Pfrieger, F.W., Bezzi, P., Bischofberger, J., et al., 2015. Synaptic integration of adult-born hippocampal neurons is locally controlled by astrocytes. Neuron 88, 957–972.
Tashiro, A., Sandler, V.M., Toni, N., Zhao, C., Gage, F.H., 2006. NMDA-receptor-mediated, cell-specific integration of new neurons in adult dentate gyrus. Nature 442, 929–933.
Tashiro, A., Makino, H., Gage, F.H., 2007. Experience-specific functional modification of the dentate gyrus through adult neurogenesis: a critical period during an immature stage. J. Neurosci. 27, 3252–3259.
Toda, T., Parylak, S.L., Linker, S.B., Gage, F.H., 2018. The role of adult hippocampal neurogenesis in brain health and disease. Mol. Psychiatry.

Tonegawa, S., Pignatelli, M., Roy, D.S., Ryan, T.J., 2015. Memory engram storage and retrieval. Curr. Opin. Neurobiol. 35, 101–109.

Tozuka, Y., Fukuda, S., Namba, T., Seki, T., Hisatsune, T., 2005. GABAergic excitation promotes neuronal differentiation in adult hippocampal progenitor cells. Neuron 47, 803–815.

Urbán, N., Guillemot, F., 2014. Neurogenesis in the embryonic and adult brain: same regulators, different roles. Front. Cell. Neurosci. 8.

Urbán, N., van den Berg, D.L.C., Forget, A., Andersen, J., Demmers, J.A.A., Hunt, C., Ayrault, O., Guillemot, F., 2016. Return to quiescence of mouse neural stem cells by degradation of a proactivation protein. Science 353, 292–295.

van Praag, H., Christie, B.R., Sejnowski, T.J., Gage, F.H., 1999. Running enhances neurogenesis, learning, and long-term potentiation in mice. Proc. Natl. Acad. Sci. U. S. A. 96, 13427–13431.

Westmark, P.R., Westmark, C.J., Wang, S., Levenson, J., O'Riordan, K.J., Burger, C., Malter, J.S., 2010. Pin1 and PKMzeta sequentially control dendritic protein synthesis. Sci. Signal. 3, ra18.

Yao, Y., Kelly, M.T., Sajikumar, S., Serrano, P., Tian, D., Bergold, P.J., Frey, J.U., Sacktor, T.C., 2008. PKM maintains late long-term potentiation by N-ethylmaleimide-sensitive factor/GluR2-dependent trafficking of postsynaptic AMPA receptors. J. Neurosci. 28, 7820–7827.

Zhang, J., Jiao, J., 2015. Molecular biomarkers for embryonic and adult neural stem cell and neurogenesis. Biomed. Res. Int. 2015, 1–14.

Zhao, C., 2006. Distinct morphological stages of dentate granule neuron maturation in the adult mouse hippocampus. J. Neurosci. 26, 3–11.

CHAPTER

7

Adult Neurogenesis in the Subventricular Zone

OUTLINE

https://doi.org/10.1016/B978-0-12-811014-0.00007-X

LEARNING OBJECTIVES

1. Describe how the neurogenic niche of the SVZ is similar and different from the niche in the dentate gyrus.
2. Discuss two signaling pathways that are important to maintaining NSCs in the SVZ.
3. Give an overview of fate specification through integration of an adult-born interneuron.
4. Describe the contributions of niche cells to proliferation and migration.
5. Explain how locomotion is accomplished by neuroblasts in the RMS.
6. Discuss the factors that affect survival of adult-born interneurons.
7. Explain how centrifugal innervation affects maturation and integration of adult-born interneurons.
8. Discuss the role of SVZ adult neurogenesis in olfactory learning and memory.

INTRODUCTION

The olfactory bulb (OB) receives new neurons throughout life that are generated in the subventricular zone (SVZ) of the lateral ventricles. In mice these structures are ~5 mm apart, and the new cells migrate from the SVZ through the rostral migratory stream (RMS) to the OB (Fig. 7.1). Approximately 10,000 new cells are born daily, but only half survive past the first 2 weeks, after which the survivors become interneurons of the OB. There are two main types of interneurons in the OB: granule cells and periglomerular (PG) cells. Of adult-born cells, 95% become granule cells, while only 5% become PG cells. Adult neurogenesis in the SVZ is thought to contribute to olfactory discrimination and learning through synaptic changes in the OB.

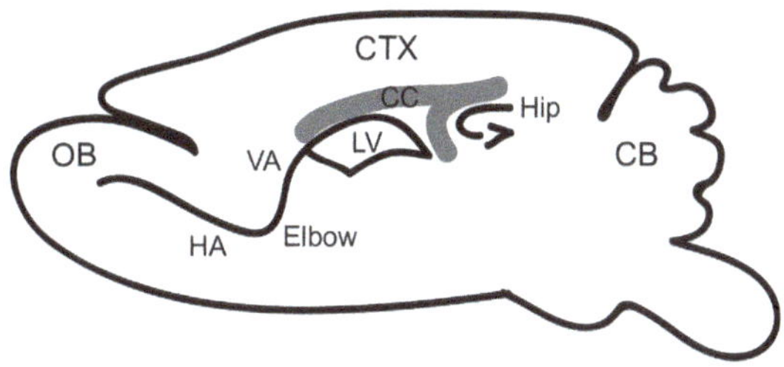

FIG. 7.1 A sagittal view of the rodent brain. Adult neurogenesis occurs in the SVZ of the lateral ventricles (LV) and dentate gyrus of the hippocampus (Hip). Neuroblasts born in the SVZ migrate ventrally through the vertical arm (VA) of the RMS, turn rostrally at the elbow, and continue migrating through the horizontal arm (HA) of the RMS to the OB.

ANATOMY OF ADULT NEUROGENESIS IN THE RODENT FOREBRAIN

The ventricular system is a series of cerebral spinal fluid (CSF)-filled cavities within the brain. The paired lateral ventricles project into the forebrain, where the SVZ neurogenic niche is found along the dorsolateral walls next to the striatum (Fig. 7.2; Mitro and Palkovits, 1981). The inside is lined with choroid plexus, a cellular layer that produces CSF (Wu et al., 1999). A single layer of ciliated ependymal cells separates the ventricles from the brain. The ependymal cilia face the ventricular side, and the beating movement of the cilia propel the CSF through the ventricles (Sawamoto et al., 2006; Yamadori and Nara, 1979). Opposed to the ependymal cells within the SVZ are the proliferating cells of the neurogenic niche.

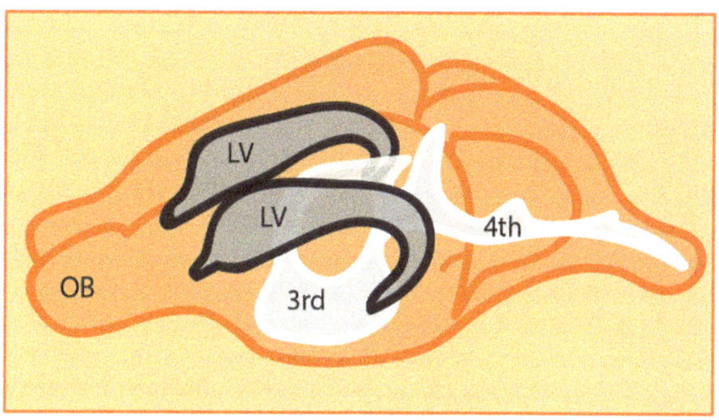

FIG. 7.2 The ventricles of the rodent brain. *LV*, lateral ventricles; *OB*, olfactory bulb; *3rd*, third ventricle; *4th*, fourth ventricle.

The Subventricular Zone

NSCs (Type B) in the SVZ give rise to intermediate neural progenitors (Type C) that give rise to migratory neuroblasts (Type A). Type B NSCs cells have large cell bodies and are usually found adjacent to ependymal cells, forming a lamina-like covering of the ependymal layer. Type B cells extend a short cilium that interdigitates between ependymal cells, such that Type B cells make direct contact with the ventricle (Fig. 7.3). The Type C intermediate neural progenitors are frequently found in contact with Type A cells and are dispersed throughout the SVZ in small clusters. Type C cells are also found in contact with either ependymal cells or the striatal parenchyma. Proliferating cells in the SVZ are arranged in such a way that Type B cells and astrocytes serve as both a path and boundary for

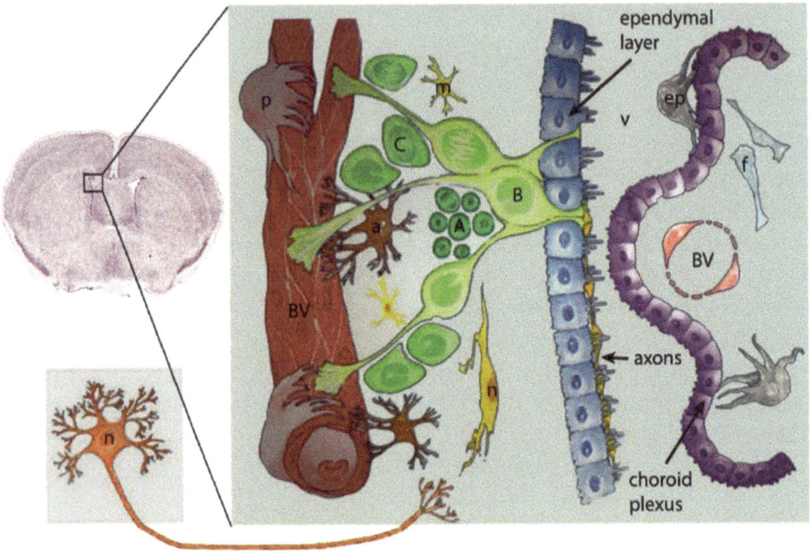

FIG. 7.3 Cells of the neurogenic niche in the adult SVZ. (Top left) The neurogenic niche occupies the dorsolateral part of the lateral ventricles. (Right) The organization of cells in and around the neurogenic niche. Pericytes (p) cover the blood vessels (BV), which are ensheathed by the endfeet of the Type B (B) NSCs. Type B cells give rise to Type C (C) cells, which give rise to Type A (A) neuroblasts. The neuroblasts use the BVs to migrate toward the RMS. Interspersed among the Type A, B and C cells are nonstem cell astrocytes (a), microglia (m) and cholinergic neurons (n, yellow). Type B cells have a single cilium that extends between the ependymal (e) cells that line the ventricle at regularly spaced intervals. The layer of choroid plexus along the e cells secretes CSF. Serotonergic neurons (axons, yellow) from the raphe extend axons throughout the ciliated side of e cells, while dopaminergic neurons (n, orange) from the substantia nigra pars compacta extend axons into the subependymal space. Epiplexus (ep) cells are macrophages that reside inside the ventricle. Other cells near the choroid plexus include blood vessel epithelia and fibroblasts (f). *Reprinted with permission from Bjornsson et al. (2015).*

the migratory Type A cells (Fig. 7.3; Doetsch et al., 1997; Lois et al., 1996). Type B1 cells prevent Type A cells from contacting the ependymal cells, while astrocytes separate Type A cells from the striatum.

The Rostral Migratory Stream

The RMS is defined by a blood vessel scaffold, on which neuroblasts migrate, ensheathed by astrocytes (Whitman et al., 2009). These tube-like structures extending from the lateral ventricles follow the rostral edge of the striatum in a rostral-ventral direction called the vertical arm, then bend (elbow region) toward and extend (horizontal arm) to the OB (Fig. 7.1). Within the RMS, migrating Type A neuroblasts form chain-like bundles directly opposed to parallel-running blood vessels (Lois et al., 1996; Whitman et al., 2009). Type A neuroblasts can halt and undergo cell division during their migration to the OB, though proliferation is altogether less than that measured in the SVZ (Luskin and Boone, 1994; Mobley et al., 2013).

The Olfactory Bulb

The OB consists of several layers, each made up of either different cell types or dendrites (Fig. 7.4). Beginning in the innermost layer, the tangentially migrating Type A neuroblasts make up the subependymal zone (SEZ). Surrounding the SEZ is the granule cell layer (GCL) consisting of interneurons that modify the output of the surrounding mitral cell layer. Mitral and tufted (mitral/tufted) cells are the projection neurons of the OB whose axons project to higher brain regions such as the amygdala and entorhinal cortex. Mitral/tufted cells have two types of dendrites—one primary dendrite that projects outwardly to the glomerular layer where it synapses with axons from neurons in the olfactory epithelium and two or more lateral dendrites that project through the external plexiform layer (EPL) and make reciprocal *dendrodendritic* synapses with granule cells. Tufted cell lateral dendrites segregate preferentially to the superficial EPL, while mitral cell lateral dendrites are mainly found in the deep EPL (Nagayama et al., 2014). In the EPL, cell bodies of short axon cells, multipolar cells and tufted cells can be found scattered throughout the granule cell dendrites and lateral dendrites of mitral/tufted cells (Nagayama et al., 2014). Granule cell activity modulates mitral/tufted cell output via *lateral inhibition* (Bartel et al., 2015). Superficial to the EPL, the glomerular layer consists of the mitral/tufted cell primary dendrites and olfactory sensory neuron axons, the source of olfactory input. Each *glomerulus* is dedicated to a population of olfactory sensory neurons expressing the same odorant receptor gene, one of more than 1000 in mice (Malnic et al., 1999). Glomeruli are surrounded by PG and external tufted cells that

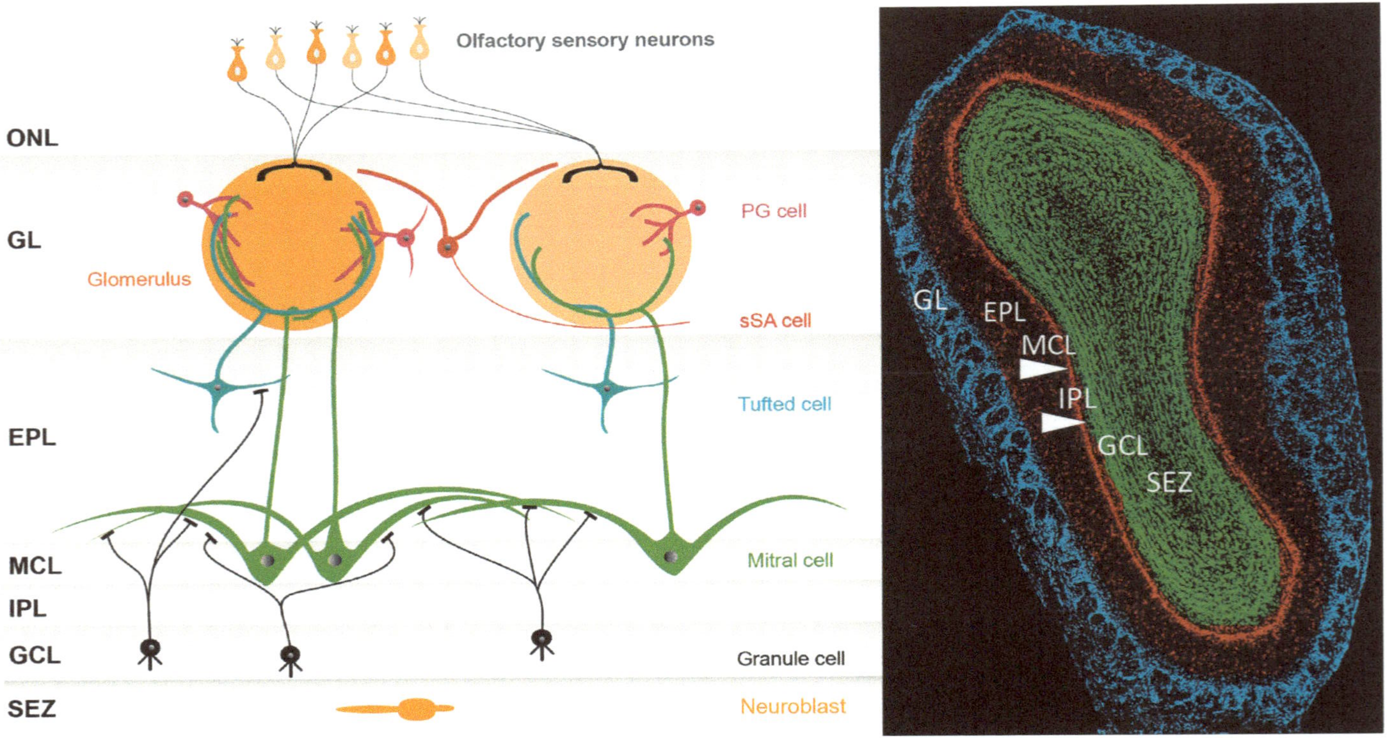

FIG. 7.4 Layers of the OB. (Left) A circuit diagram of the connections between cells of the OB and olfactory sensory neurons in the nose. *ONL*, olfactory nerve layer; *GL*, glomerular layer; *EPL*, external plexiform layer; *MCL*, mitral cell layer; *IPL*, internal plexiform layer; *GCL*, granule cell layer; *SEZ*, subependymal zone; *sSA*, superficial short axon cell. (Right) A fluorescently labeled image of a coronal section through the OB. *(Left) Image modified from Nagayama, S., Homma, R., Imamura, F., 2014. Neuronal organization of olfactory bulb circuits. Front. Neural Circ. 8; under the terms of the Creative Commons Attribution License (CC BY). Copyright © 2014 Nagayama, Homma and Imamura. (Right) Image by Matt Valley, available in the public domain.*

modify mitral cell output, creating a circular structure of cell bodies on the periphery and synaptic connections within each glomerulus (Nagayama et al., 2014). The outermost layer is the olfactory nerve layer (ONL) that consists of the olfactory sensory neurons axons from the olfactory epithelium in the nose.

Once Type A neuroblasts reach the OB they enter the SEZ (Lois et al., 1996). The direction of migration shifts from tangential to radial migration, targeting one of two layers, the GCL or glomerular layer (Box 7.1). The GCL can be divided into deep and superficial sublayers. Most adult-born neuroblasts (95%) will integrate into the deep GCL and become granule cells (Lemasson, 2005; Lledo and Saghatelyan, 2005). The remaining 5% of neuroblasts migrate to the glomerular layer and become PG cells (Lledo and Saghatelyan, 2005).

BOX 7.1

SPOTLIGHT ON MIGRATION

The Type A neuroblasts form an extensive network of chains that converge at the anterior SVZ and extend all the way to the SEZ of the OB. Chain migration, also described as tangential, is a term based on neural development and the orientation of migrating cells to the radial glia (Cooper, 2013). Once in the OB, neuroblasts detach from the chains and migrate radially.

Neuronal migration, whether it is tangential or radial requires initiation, guidance, and locomotion, ending with detachment. Slit molecules initiate neuroblast repulsion from the SVZ, orienting them toward the OB. Guidance is provided by BDNF, secreted by the endothelial cells and Netrin-1 in the OB. Changes to the cytoskeleton and cell adhesion molecule expression are required to carry out the stages of migration.

The cytoskeleton is made up of microtubules and actin filaments. Microtubule dynamics underlie the ability of a cell to migrate. Microtubules are composed of α–β polymers that originate from a specialized structure in the cell called the centrosome, or organization center. From the centrosome microtubules extend by the addition of single polymers. The end that exposes the α-tubulin end is called the minus end, where slow growth occurs, while the β-tubulin side is the plus end where rapid growth occurs. Actin filaments are made up of actin subunits that have a barbed and pointed end, with filament growth typically occurring at the barbed end.

Cell adhesion molecules enable the Type A neuroblasts to move through the different environments of the SVZ, RMS and OB. NCAMs are a family of molecules that can be found throughout the nervous system.

Continued

BOX 7.1

SPOTLIGHT ON MIGRATION—CONT'D

They are expressed on the cell surface and facilitate cell-cell interactions typically by forming homophilic interactions with NCAM molecules expressed by neighboring cells. PSA-NCAM is specifically expressed by Type A neuroblasts and cannot be substituted with NCAM. Removal of PSA results in disruption of chains and scattered migration of the neuroblasts (Battista and Rutishauser, 2010).

Matrix metalloproteases (MMPs) are zinc-dependent proteases that can modify the ECM, a complex network of secreted proteins and polysaccharides found outside of and in between the cells. Often cell interactions with the ECM must be modified in order for migration to proceed. MMPs are secreted by endothelial cells in the SVZ and promote neuroblast migration, though they interact with individual cells and do not affect neuroblast migration when they are in chains (Bovetti et al., 2007; Wang, 2006).

The integrins are another large family of *glycoproteins*, whose members mainly form heterodimers consisting of one α and one β subunit, to connect cell surface molecules with the ECM. The α6β1 integrin is expressed by migrating neuroblasts that interact with the ECM protein laminin (Emsley, 2003; Jacques et al., 1998). Interestingly, Type B NSCs also express α6β1 integrin, but at higher levels than the neuroblasts, altering the adhesiveness between these populations (Shen et al., 2008). Laminin surrounds the blood vessels and is important for physically guiding neuroblasts through the RMS (Emsley, 2003).

A migrating neuroblast goes through four main stages to create locomotion along the scaffold provided by the blood vessels and astrocytes in the RMS. First, the leading process extends by microtubule growth. Next, the centrosome, which lies between the nucleus and the leading process, is moved forward. Microtubules that extend from the centrosome bring the nucleus forward, wrapping around it like a cage. Nucleus movement is called nucleokinesis. The last step is remodeling of the trailing process to bring it closer to the cell body. The *motor protein* myosin II is highly concentrated at the rear of the cell where it contracts actin, which acts to push the nucleus forward (Schaar and McConnell, 2005; Vallee et al., 2009).

Once in the OB, a gradient of Reelin, most strongly secreted by mitral cells, signals detachment from neuroblast chains (Hack et al., 2002; Valero et al., 2007). Neuroblasts switch from tangential to individual radial migration through the OB by downregulating Sphingosine 1-phosphate (S1P). S1P is a bioactive lipid secreted by astrocytes that interacts with the sphingosine 1-phosphate receptor 1 (S1P1) on neuroblasts, promoting tangential neuroblast migration through the RMS (Alfonso et al., 2015). Thus multiple signaling pathways regulate each phase of neuroblast migration from the SVZ to the OB.

PHASES OF ADULT-BORN CELL MATURATION

The phases of maturation for cells born in the adult SVZ often occur at a distance from the previous phase and are thus less influenced by changes. For example, removal of the OB does not impact proliferation and migration in the RMS for more than 3 weeks (Kirschenbaum et al., 1999). Thus factors of migration, survival, maturation and integration in the RMS and OB are less likely to affect maintenance, proliferation and fate specification in the SVZ.

Phase 1: Stem Cell Maintenance, Activation and Fate Specification

Type B NSCs are multipotent and give rise to oligodendrocytes, neurons, and astrocytes (Doetsch et al., 1999; Gritti et al., 1999). While all Type B NSCs express the glial marker, glial fibrillary acidic protein (GFAP), the quiescent Type B cells can be differentiated from active Type B cells by Id1, a helix-loop-helix transcriptional regulator expressed in the active (proliferating) cells (Nam and Benezra, 2009). Anti-mitotic agents infused into the SVZ cause a loss of proliferating cells, including Id1-positive Type B cells and Type C cells. Subsequently, quiescent Type B cells become active and restore the Type C cells (Doetsch et al., 1999). Proliferating Type B and Type C cells can be differentiated by GFAP and epidermal growth factor (EGF), where Type B cells are GFAP- and EGF-positive (Pastrana et al., 2009). Through asymmetric cell division, Type B cells give rise to GFAP-negative/EGF-positive Type C cells (Pastrana et al., 2009).

The fate of the adult-born cells in the OB is determined by the location of the stem cell that gave birth to it in the SVZ/RMS. Granule cell interneurons in the OB all express the inhibitory neurotransmitter GABA, with some expressing the calcium binding protein calretinin (Kohwi et al., 2007). Within the PG cell population the majority express GABA, but there are three nonoverlapping subtypes characterized by their expression of the dopamine precursor *tyrosine hydroxylase* (TH), calretinin, and another calcium binding protein, calbindin (Lledo et al., 2008; Whitman and Greer, 2007a). In the mouse OB, the TH-positive and calbindin-positive PG cells express GABA, while a majority (65%), but not all of the calretinin-positive PG cells also express GABA (Kosaka and Kosaka, 2007).

There are three main divisions of the SVZ, which are defined by the types of interneurons produced in each (Fig. 7.5). The divisions were first reported in a study targeting SVZ stem cells by injecting an adenovirus reporter into the ventricle (Merkle et al., 2007). The most dorsal region of the SVZ contains progenitors that give rise to dopaminergic (TH-positive) PG cells and superficial granule cells (Young et al., 2007). Within the dorsal region, the transcription factor empty spiracles homeobox 1 (Emx1)-positive

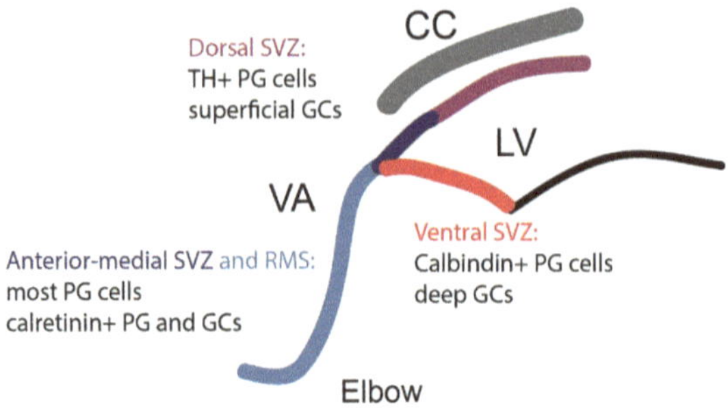

FIG. 7.5 The divisions of the SVZ based on interneuron production.

progenitors give rise to calretinin and dopaminergic PG cells (Kohwi et al., 2007). In the most ventral SVZ, progenitors expressing the transcription factor Nkx2.1 give rise to calbindin-positive PG cells and deep granule cells (Merkle et al., 2007; Young et al., 2007). The RMS and medial septal wall gave rise to calretinin-positive granule cells and PG cells (Kohwi et al., 2007). Across the SVZ and RMS, the medial septal wall produces the largest number of interneurons (Merkle et al., 2007).

Phase 2: Expansion of Intermediate Neural Progenitors

Type C cells were determined to be proliferating progenitors based on morphology, proliferating activity and proximity to Type A and Type B cells (Doetsch et al., 1997). Type C cells have a cell cycle of 18–25h with more than half of that time spent in S-phase (14–17h; Ponti et al., 2013). Type C cells divide symmetrically three times before differentiating into neuroblasts (Ponti et al., 2013). Confirmation that Type A cells are proliferating, migrating neuroblasts that give rise to neurons, was shown by the use of [3H]thymidine autoradiography or BrdU immunohistochemistry (Doetsch et al., 1997). Type A cells make up approximately half of the dividing cells in the SVZ and divide symmetrically once or twice before migrating into the RMS (Ponti et al., 2013; Saha et al., 2012).

Phase 3: Migration and Initial Pruning of Newborn Granule Cells

Type A neuroblasts migrate approximately 5mm from the SVZ, through the RMS to the OB in the mouse brain (Box 7.1; Alvarez-Buylla and Garcia-Verdugo, 2002). The Type A cell has an elongated cell body with a short trailing process and an extended leading process (Lois et al., 1996). The rate of neuroblast migration in the RMS was reported by one group to be 23μm/h (Luskin and Boone, 1994). A later study that used live imaging of brain slices labeled with DiI (1,1′-dioctadecyl-3,3,3′,3′-tetramethylindocarbocyanine perchlorate; pronounced dye-eye) reported a neuroblast migration rate between ~80 and 100 μm/h (Murase and Horwitz, 2002). The discrepancy between the two studies may be due to neuroblasts pausing in migration to proliferate. Neuroblasts reach their target 2–7 days after leaving the SVZ (Petreanu and Alvarez-Buylla, 2002). It is estimated that approximately 10,000 neuroblasts migrate to the OB per day, though half of them die within the first 4 weeks (Petreanu and Alvarez-Buylla, 2002). Studies suggest that olfactory-relevant behaviors influence the survival and integration of Type A neuroblasts (Bastien-Dionne et al., 2010; Bovetti et al., 2009; Lemasson, 2005; Lledo and Saghatelyan, 2005). Damage to proliferating cells by ionizing radiation also increases survival, which is thought to be a compensatory mechanism in response to a reduction in new cells being produced (Díaz et al., 2017).

Phase 4: Maturation and Functional Integration of Newborn Neurons

The first task of a neuroblast in the OB is to migrate radially to its new location. After reaching the target layer, morphological and electrophysiological differentiation proceeds for 14–28 days, which includes dendritic spine formation (Whitman and Greer, 2007b). The density of dendritic spines increases significantly for another 2 weeks, followed by a decrease over another 2 weeks for a total of 56 days to mature (Whitman and Greer, 2007b).

The differentiation process of a granule cell interneuron in the OB can be divided into five stages based on phenotype and electrophysiology (Carleton et al., 2003; Petreanu and Alvarez-Buylla, 2002). Beginning with stage 1, cells in the SEZ have round cell bodies each with a concise leading process and a growth cone consistent with the morphology of a migrating cell (Petreanu and Alvarez-Buylla, 2002). At stage 2, growth cones are no longer visible, and the cell bodies are larger with a single process that extends toward the mitral cell layer in a radial orientation. Radial migration is complete in stage 3; there is no longer a trailing process, and the cell bodies are settled in the GCL next to mature granule cells. Neurites begin developing from the single process on the basal aspect, extending into the GCL. At stages 4 and 5, more complex branched apical dendrites extend beyond the mitral cell layer. Stage 5 is differentiated from stage 4 by the presence of dendritic spines (Petreanu and Alvarez-Buylla, 2002).

A mature interneuron is bipolar, but axonless; its dendrites can be divided into three domains. The basal domain is the basal dendrite oriented toward the SEZ (Fig. 7.4). The proximal domain is the initial apical dendritic segment oriented toward the EPL. The distal domain begins at the first branch point of the apical dendrite. Synapses are first acquired in the proximal domain, forming glutamatergic synapses with axons from the olfactory cortices (Bardy et al., 2010; Kelsch et al., 2008; Soria-Gómez et al., 2014; Whitman and Greer, 2007b). The next synapses acquired are dendrodendritic connections formed with mitral cells in the distal domain (Hinds and Hinds, 1976). Mitral cells release glutamate that activates the release of GABA from the granule cell back onto its network of mitral cells. This is called lateral inhibition and refines the number of mitral/tufted cells responding to a single odor (Najac et al., 2015). The number of synapses in the dendritic domain reflect the size of its network and number of projection neurons receiving inhibition from the granule cell (Lepousez et al., 2013).

Electrophysiologically, adult-born granule cells undergo changes in membrane properties and synaptic activity that reflect the developing morphology of synapse production (Carleton et al., 2003). During neuroblast migration, the cells already express $GABA_A$ and AMPA receptors, similar to developing embryonic neurons (Stewart et al., 2002). Cells with stage 1 and 2 morphology express voltage-gated K+ channels, without evidence of voltage-gated Na+ channels or spontaneous synaptic currents. AMPA receptor currents are present in a small subset of Class 1 and most Class 2 cells, though only 50% of Class 2 cells express NMDA receptors (Carleton et al., 2003). Despite the fact that GABA receptor activation produces IPSCs, and AMPA receptor activation produces EPSPs in Class 3 and Class 4 cells, they are still unable to fire action potentials. A separate study confirmed that adult-born granule cells receive synaptic input on their cell bodies before their dendritic arbors integrate into the existing circuitry in the EPL (Whitman and Greer, 2007b). Drs. Whitman and Greer have described the process as "listening before speaking." The Class 5 cells have properties similar to the surrounding mature granule cells. Analysis of EPSPs confirmed two distinct glutamatergic sources, consistent with morphological observation. The first glutamatergic synapses are with mitral/tufted cell axons or *centrifugal* fibers, then with the lateral dendrites of mitral/tufted cells (Carleton et al., 2003).

Survival of adult-born interneurons in the OB is closely tied to maturation. Accumulating evidence suggests that new neurons that are unable to complete any step in maturation are eliminated (Kelsch et al., 2009, 2012; Yamaguchi and Mori, 2005). Cell-intrinsic activity, such as voltage-gated K+ and Na+ channel current, is a major determinant of survival (Lin et al., 2010). The chance of survival is also increased by neural activity due to olfactory behavior. One study found that, similar to immature neurons in the dentate gyrus, operant conditioning resulted in greater survival rates of SVZ adult-born neurons than in nonoperant learning trials (Mandairon et al., 2011). Thus survival is not so much a phase as a possibility.

THE SVZ NICHE

The SVZ niche is similar to the dentate gyrus in that it supports the maintenance and proliferation of neural progenitor cells. It differs in that fate specification (SVZ), migration (RMS) and survival and integration (OB) occur in different brain regions. Fate specification is also more complex in that there are multiple neuron types into which an adult-born SVZ could differentiate. The surrounding cells, astrocytes, microglia, blood vessels, pericytes and ependymal cells, influence the behaviors that occur within each subniche. NSCs transplanted into nonneurogenic regions show limited neurogenesis, demonstrating the importance of the niche to NSC biology (Temple, 2001). After a stroke, neuroblasts from the SVZ will readily migrate to the damaged area, but most become glia (Gotts and Chesselet, 2005; Parent et al., 2002). Thus an understanding of how the niche supports neuronal differentiation and survival has clinical relevance.

Ependymal Cells

Type B cells extend an apical cilium into the center of pinwheel-like clusters of ependymal cells (Mirzadeh et al., 2008). The close association of Type B and ependymal cells induces the pinwheel architecture, as this ependymal organization is only observed over the neurogenic niche (Mirzadeh et al., 2008). Type B cells share apical junctional complexes (gap and adherens junctions) between other Type B cells and ependymal cells that could be important for determining NSC symmetric and asymmetric division (Chenn and Walsh, 2002; Kosodo et al., 2004). Ependymal cells also produce Noggin, a secreted protein that inactivates BMPs and increases Type B cell neurogenesis (Lim et al., 2000; Peretto et al., 2004). Within the ventricle, the NSC cilia are able to detect signals within the CSF. The choroid plexus produces basic fibroblast growth factor (bFGF) that enhances adult neurogenesis. The CSF also contains insulin-like growth factor 1 (IGF1), another factor known to increase adult neurogenesis (Lehtinen et al., 2011; Sun et al., 2009). Thus the organization and association of ependymal cells is an important part of the SVZ niche.

Endothelial Cells

Type B cells have a basal process with a large endfoot that ensheaths blood vessels (endothelial cells) in the SVZ (Tavazoie et al., 2008). Endothelial cells influence the behavior of Type B cells through physical interaction, expression and secretion of signaling molecules. Physically, the endothelial cells can regulate NSC proliferation through adhesion to the extracellular matrix (ECM) protein laminin, a multisubunit protein that can bind to both ECM components and laminin receptors expressed by cells (Box 7.1). Type B cells express the laminin receptor $\alpha6\beta1$, through which there is adhesion to the endothelial cells (Shen et al., 2008). Tighter adhesion reduces proliferation, while blocking adhesion increases proliferation (Shen et al., 2008). By signaling through the Notch pathway (below), endothelial cells promote self-renewal and neurogenesis (Shen, 2004). Finally, endothelial cells also produce soluble factors such as pigment epithelium-derived factor (PEDF) that enhances Notch signaling in activated Type B cells, and results in symmetric, rather than asymmetric proliferation (Andreu-Agulló et al., 2009).

In the RMS, endothelial cells release brain-derived neurotrophic factor (BDNF), which binds to the neurotrophin receptor p75 expressed by neuroblasts, and regulates neuroblast migration (Snapyan et al., 2009). Thus neuroblasts are guided by BDNF-secreting blood vessel scaffolds into the OB.

Astrocytes

Nonstem cell astrocytes are abundant in the niche and contribute in important ways to adult neurogenesis. In the SVZ, astrocytes secrete a Notch ligand, Dlk1, that promotes neurogenesis (Ferrón et al., 2011). In the RMS, astrocytes have important developmental and postnatal roles. During development VEGF-expressing astrocytes guide the development of the vasculature in the RMS, without which neuroblast migration is compromised (Bozoyan et al., 2012). Postnatally, astrocytes regulate neuroblast migration in two different ways. First, expression of Thrombospondin 4 (Thbs4) is important for normal neuroblast migration, specifically for the calbindin- and parvalbumin-expressing populations (Girard et al., 2014). Second, the release of GABA from astrocytes reduces the mean rate of neuroblast migration through activation of $GABA_A$ receptors, from ~50 μm/h to ~30 μm/h (Bolteus and Bordey, 2004). In the OB, astrocytes respond to the activation of glutamate receptors on principal neurons by increasing blood flow to support intense activity, and thereby likely contribute to the survival of integrating neurons in the OB (Takano et al., 2006). Thus throughout the neurogenic forebrain, nonstem cell astrocytes play active roles in adult neurogenesis.

Microglia

Microglia in the SVZ, RMS, and OB have distinct roles and morphological properties compared to microglia in the cortex, suggesting an adaptation to the unique physiology of these regions. In the cortex, microglia have a ramified appearance with multiple thin processes extending from the cell body (Fig. 7.6). When in a ramified state, they actively survey the surrounding microenvironment by moving their processes and migrating through the CNS (Karperien et al., 2013; Nimmerjahn, 2005). In the SVZ, RMS, and OB they are more amoeboid-like, with large cells bodies and only a few thick processes (Ribeiro Xavier et al., 2015). Amoeboid microglia are also referred to as reactive or unramified. In this state they have an active immunoinflammatory role in which they travel to sites of

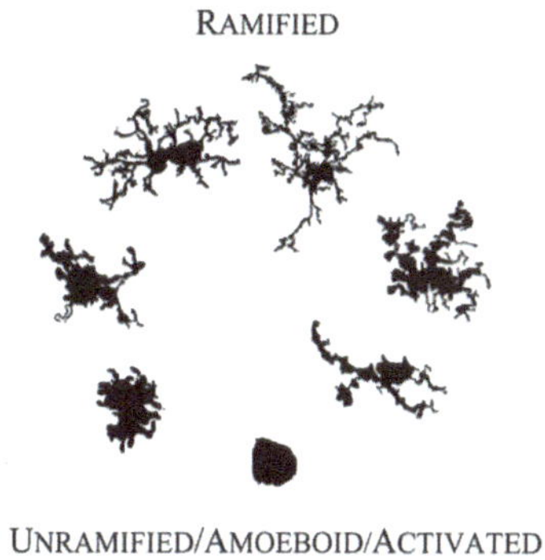

FIG. 7.6 Metamorphosing microglia. In a ramified state (top), microglia have branches with a feathery or spiny appearance. Reactive microglia (bottom, moving clockwise) have a larger cell body and thicker processes. *Reprinted under the terms of the Creative Commons Attribution License. Copyright © 2013 Karperien, Ahammer and Jelinek.*

injury or inflammation where they may recruit immune cells or clear dead or dying cells by phagocytosis (Karperien et al., 2013). Microglia in the SVZ and RMS do not engage in phagocytic activity, and in vitro, microglia secrete factors that promote the production of neuroblasts by Type B NSCs (Walton et al., 2006). In the RMS, microglia are important for neuroblast survival and unidirectional migration (Ribeiro Xavier et al., 2015). Conversely, in the OB a majority of microglia are phagocytic, likely eliminating neuroblasts pruned during the first 14–21 days in the OB.

The unique qualities of microglia within the neurogenic forebrain is a subject of interest for researchers interested in neuroregeneration (Xiong et al., 2016). For example, following a stroke, neuroblasts readily migrate to the damaged area, but most differentiate into astrocytes. If microglia of the SVZ and RMS were also in the damaged region, their presence might promote neurogenesis and improve the outcome of poststroke recovery.

REGULATING THE BEHAVIORS OF ADULT-BORN CELLS

NSC Maintenance

The Delta/Notch signaling pathway has an important role in maintaining the Type B NSC population by antagonizing proneural genes and neuronal differentiation. When the Notch receptor is activated by its ligands, the Notch intracellular domain (NICD) translocates to the nucleus where, in mammals, it forms a complex with recombinant binding protein j (RBPJ; Imayoshi and Kageyama, 2011). The NICD-RBPJ complex induces expression of the transcriptional repressor genes *Hes1* and *Hes5.*

The two Hes genes are responsible for repressing proneural genes, inhibiting differentiation and maintaining an ability to self-renew (Imayoshi and Kageyama, 2011). Two components of the niche contribute to Notch signaling regulation: endothelial cells and Type C cells. Endothelial cells of blood vessels express two Notch ligands, Jag1 and Dll4 (High et al., 2008). Jag1 contributes to the maintenance of quiescent Type B cells (Ottone and Parrinello, 2015). The Type C cells express the EGF receptor, which suppresses Notch signaling (Aguirre et al., 2010). This interaction between Notch and EGF-receptor signaling maintains a balance between the NSC and neuroprogenitor cell pools (Aguirre et al., 2010). Thus Notch signaling is essential for stem cell maintenance.

Sonic hedgehog (Shh) is a morphogen protein, responsible for patterning events during development (Ericson et al., 1996; Litingtung and Chiang, 2000). Shh binds a membrane-bound receptor called Patched resulting in its down-regulation (Fig. 7.7). In the absence of Patched, another transmembrane-bound protein Smoothened moves to the cell surface where it can inhibit the cleavage of a GLI family transcription factor. Intact GLI is a transcriptional activator, and the GLI family member Gli1 is expressed in the ventral SVZ. Shh signaling activates genes that shorten the cell cycle of active Type B cells, resulting in the progressive ex-

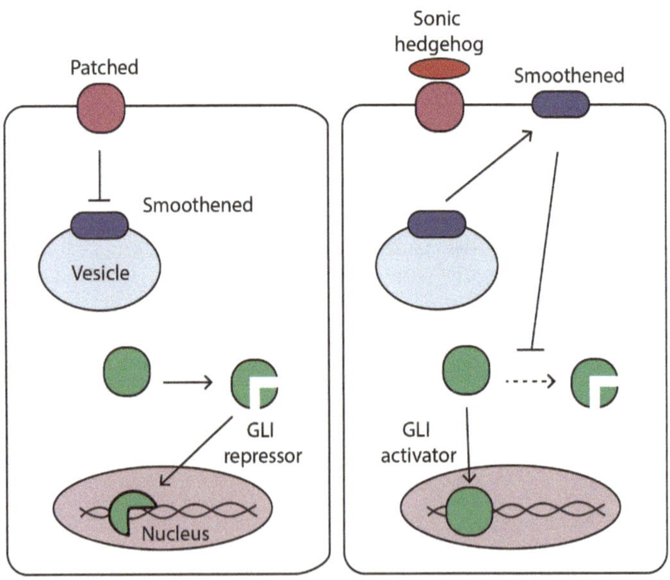

FIG. 7.7 Sonic hedgehog signaling. Patched is the receptor for Sonic hedgehog (Shh) and is found in the membrane of cells. Patched inhibits Smoothened from insertion in the surface membrane until Shh binds to Patched. The GLI family transcription factor is cleaved by a protease complex that is a transcriptional repressor. Once Smoothened is in the surface membrane, it inhibits GLI cleavage, converting GLI to a transcriptional activator.

haustion of active cells, while increasing the number of quiescent Type B cells (Daynac et al., 2016). As such, Shh signaling has an important role in the adult SVZ as it balances the number of active and quiescent stem cells (Daynac et al., 2016).

Fate Specification

The fate choices reviewed for stem cells in the dentate gyrus apply to Type B cells in the SVZ. Additionally, Type B cells give rise to different populations of interneurons in the OB. Shh signaling also plays an important role in Type B cell fate specification. Shh is secreted from neurons in the adult striatum and determines transcription factor expression in ventral SVZ stem cells through the Smoothened protein (Ihrie et al., 2011). Transfecting dorsal stem cells with the Smoothened protein reprograms the cells to produce ventral-derived progeny (Ihrie et al., 2011). The ability to reprogram the fate of adult-born neurons has important therapeutic relevance. For example, reprogramming and targeting adult-born cells to replace neurons lost to degeneration due to Alzheimer or Parkinson disease is a topic that is being investigated.

One of the fate choices for Type B cells is to give rise to oligodendrocytes. Type B NSCs give rise to Type C cells expressing Olig2, a marker of oligodendrocytes (Menn et al., 2006). The oligodendrocyte precursor cells (OPCs) born in the SVZ have the ability to migrate along the corpus collosum, striatum, and fimbria fornix and differentiate into mature oligodendrocytes (Menn et al., 2006). Prox-1 is a transcription factor important for cell cycle regulation, and in the adult dentate gyrus, is required for neural differentiation (Karalay et al., 2011). In the SVZ, Prox-1 expression begins during the transition from Type C to a migratory cell, but is maintained only in cells destined to become oligodendrocytes (Fig. 7.8; Bunk et al., 2016). Prox-1 is an important fate determinant, but additional work is required to understand how its role in neurogenesis versus oligodendrogenesis is regulated.

Activation of Quiescent Cells

Activation of Type B cells is influenced by neuronal activity through different neurotransmitters (Fig. 7.8). Serotonergic axons from the raphe nuclei form an extensive network along the ependymal cells inside the ventricle, appearing to snake through the cilia (Tong et al., 2014). Release of serotonin onto the Type B cells increases symmetric proliferation (Tong et al., 2014). Dopaminergic neurons from the substantia nigra pars compacta also project axons to the SVZ. Dopamine can increase proliferation, though the full effect of dopamine on adult neurogenesis in the SVZ is controversial (Berg et al., 2013). A population of unique cholinergic

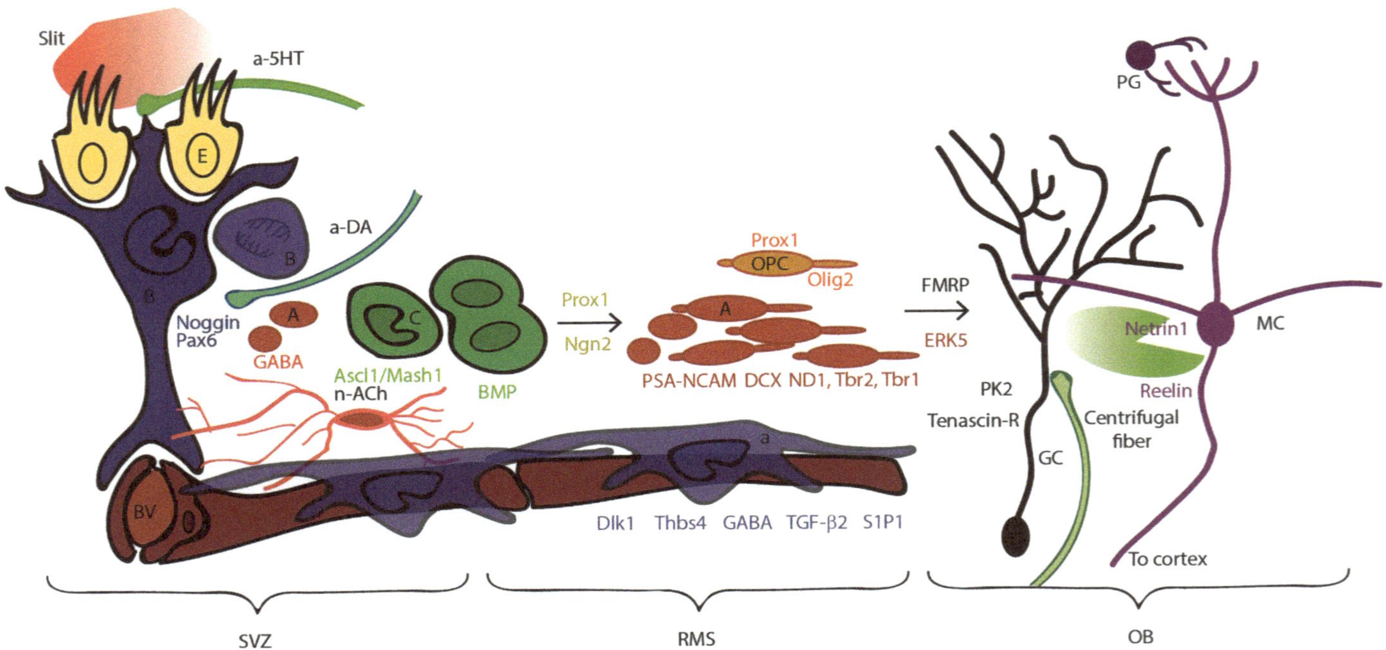

FIG. 7.8 Factors affecting adult neurogenesis. Type B (B) cells in the SVZ have an endfoot that ensheaths blood vessels (BV) and an apical cilium that protrudes into the ventricle. Type B cells divide to self-renew and/or give rise to nonstem cell astrocytes (a), or Type C (C) cells, which give rise to Type A (A) and OPC cells. Ependymal (E) cells that line the ventricle are closely opposed to Type B cells. Serotonergic (5HT) axons (a) in the ventricle, cholinergic neurons (n-ACh) and dopaminergic axons (a-DA) affect Type B cell proliferation. Slit produced by choroid plexus (see Fig. 7.3) in the ventricle is a repulsive signal to Type A cells that initiates their migration. Type A cells use blood vessels and astrocytes to migrate to the OB, attracted by Netrin-1 released by the mitral cells (MC). In the OB Type A cells differentiate into granule cells (GC) or PG cells. GC activity is modified by centrifugal fibers from other cortical areas. All other labels are factors expressed by or secreted by cells of the corresponding color or in that area.

neurons are located in the niche whose axons project locally, paralleling the ependymal cells lining the ventricle (Paez-Gonzalez et al., 2014). Release of acetylcholine results in an increase in Type B cell proliferation. Type A neuroblasts can inhibit proliferation by releasing GABA in a non-synaptic manner that binds to $GABA_A$ receptors on the Type B cells (Liu et al., 2005). As such, the neuroblast population provides feedback to the Type B NSCs to control the balance between serotonergic-activated self-renewal and GABAergic-induced quiescence (Liu et al., 2005).

Proliferation of Intermediate Neural Progenitors

The balance between Type B NSCs and Type C intermediate neural progenitors is mediated by BMP signaling (Chapter 3) and its inhibitor Noggin. BMPs are expressed by Type C cells, Type B NSCs, and endothelial cells (Mathieu et al., 2008). Noggin is a secreted polypeptide that binds and inactivates BMPs. Noggin is produced by Type B cells and the ependymal cells that line the ventricle (Lim et al., 2000; Peretto et al., 2004). Similar to the effect of BMP in the hair follicle, BMPs reduce proliferation in the Type B cells in culture (Mathieu et al., 2008). However, in vivo, BMP signaling is necessary for neurogenesis and when absent, increases oligodendrocyte production (Colak et al., 2008). Indeed, overexpression of Noggin increases the differentiation of Type B cells into Type C and OPCs, while reducing the number of astrocytes produced (Morell et al., 2015). These studies demonstrate the close interaction between SVZ niche cells and suggest a complex interplay between cells of the niche to regulate the population of proliferative cells.

Differentiation

Many proteins play multiple roles in the brain, such that expression at the right place and time are crucial for normal development and physiology. The sequence of protein expression not only ensures normal development but also the final fate of the cell. Transcription factors are a good example of proteins with precise timing that affect the differentiation of neurons. Pax6 is a transcription factor responsible for a subset of neuronal proteins involved in dendrite development (Herold et al., 2011; Kohwi et al., 2007). Pax6 expressing Type B cells give rise to Type C cells that turn on expression of achaete-scute homolog 1 (Ascl1/Mash1). There are three subpopulations of Pax6/Ascl1-positive cells: one expresses Olig2 that gives rise to oligodendrocytes; another expresses Neurogenin2 (Ngn2), a basic helix-loop-helix (bHLH) transcription factor; and the third is negative for Olig2 and Ngn2 (Roybon et al., 2009). The Ascl1/Ngn2-positive cells are limited to the caudal RMS and dorsal/dorsal-caudal SVZ. As Ascl1/Ngn2-positive cells reach the elbow of the RMS, Ascl1 expression is downregulated and they begin to express T-box brain protein 2 (Tbr2), followed by Tbr1. During the period that Tbr2 is expressed, neuroblasts downregulate Ngn2 expression and turn on another bHLH transcription factor, NeuroD1 (ND1). With each successive step in transcription factor expression, these cells are less likely to proliferate (Roybon et al., 2009). Pax6/Ascl1-positive cells show more proliferation with Tbr1-positive cells showing the least amount of proliferation. Lack of proliferation is correlated with expression of the migratory markers Polysialic Acid Neural Cell Adhesion Molecule (PSA-NCAM) and Doublecortin (DCX). The majority of these cells that survive differentiate into calretinin-positive periglomerular cells, while a minority (6%–13%) differentiate into calbindin neurons.

Studies that examine the role of proteins known to be important for neuronal differentiation often use knockout models. However, proteins that affect adult-born neurons' ability to integrate into the OB are frequently necessary for survival, and it is difficult to ascertain their role in differentiation when they are absent. Experiments in culture can help determine the role of a protein in this process. For example, ERK5 is a member of the mitogen-activated protein (MAP) kinase family that, when deleted, causes a significant decrease in surviving neuroblasts (Li et al., 2013). In cultured neuroblasts, ERK5 regulates the dendritic length, including number of branches in OB interneurons, an important step in integration (Li et al., 2013).

The role of Fragile X Mental Retardation Protein (FMRP) is specific to dendritic spine formation and affects parameters such as spine density, morphology, and activity-dependent remodeling. In neurons where FMRP expression is prevented, there is a significant increase in spine density, but they display an immature morphology (Scotto-Lomassese et al., 2011). Despite the immature dendritic spines, synapses with glutamatergic neurons still form normally. The dendritic arbors of OB interneurons respond with changes in complexity to activity levels (Saghatelyan et al., 2005). Naris closure will cause the total dendritic tree length of the interneurons to decrease, a response that is dependent on the presence of FMRP (Scotto-Lomassese et al., 2011). Interestingly, the effects of FMRP are independent of survival.

Migration

The tangential migration of neuroblasts is facilitated by adhesion molecules and guidance molecules (Box 7.1). Many components come together to ensure that neuroblasts leave the SVZ, migrate through the RMS and target the OB. Among other factors, these components work together to affect the morphology, direction, and rate of neuroblast migration through the RMS.

The *chemorepulsive* molecule, Slit is expressed by the adult septum and choroid plexus (Wu et al., 1999). The septum is the region between the two lateral ventricle protrusions into the forebrain. Slit expression in the septum and the choroid plexus repel neuroblasts from the SVZ (Wu et al., 1999). The flow of CSF, propelled by the beating of ependymal cell cilia also guides the directional migration of neuroblasts (Sawamoto et al., 2006). Disrupting CSF flow or knocking out Slit protein expression results in disorganized neuroblasts that migrate in tangential directions to their normal path to the OB resulting in a significant reduction in the number of neuroblasts that reach the OB (Nguyen-Ba-Charvet et al., 2004; Sawamoto et al., 2006).

In the RMS, endothelial blood vessel cells secrete BDNF that induces neuroblast migration, without which the neuroblasts migrate more slowly (Chiaramello et al., 2007; Snapyan et al., 2009). The neuroblasts express PSA-NCAM, a marker of Type 3 cells in the dentate gyrus, which is necessary for neuroblast migration. Without PSA-NCAM, neuroblasts aggregate in large clusters along the length, especially the caudal-most end of the RMS near the SVZ (Chazal et al., 2000). PSA-NCAM maintains the bundles of neuroblast while allowing them migratory movement (Lois et al., 1996; Rousselot et al., 1995; Rutishauser et al., 1985).

Netrin-1 is a diffusible neuron guidance molecule secreted from projection neurons in the OB. A Netrin-1 gradient provides *chemoattractive* directional guidance to the neuroblasts (Murase and Horwitz, 2002). Disruption of the Netrin-1 receptor, Deleted in Colorectal Cancer (DCC) results in a significant reduction in the rate of neuroblast migration with the leading process often directed away from the OB (Murase and Horwitz, 2002). Prokineticin2 (PK2) is another secreted molecule in the SEZ of the OB that attracts the neuroblasts from the SVZ. Once in the OB, PK2 facilitates radial migration to the granule and glomerular layers by signaling neuroblast detachment from the chains of cells entering the SEZ (Ng, 2005).

Survival

Studies on the survival of OB interneurons fall into two main categories: the molecular mechanisms of survival or the influence of olfactory behavior on survival. In the molecular category, the expression of the transcription factor Pax6 regulates the survival of dopaminergic adult-born interneurons in the OB (Ninkovic et al., 2010). The effect of Pax6 on survival requires the activity of cAMP response element-binding protein (CREB). Loss of CREB decreased interneuron survival, while remaining postmigratory neurons showed defects in dendritic arbor development. Furthermore, CREB is activated by cAMP, a small molecule produced by type 3 adenylyl cyclase (AC3) that is expressed in granule cells (Luo et al., 2015). Loss of AC3 reduced interneuron survival, and further investigation found that adult-born neurons did not develop dendritic spines. Thus Pax6-induced cAMP signaling is important for maturation, directly affecting interneuron survival. The molecules in the signaling pathway provide several different steps at which neuronal survival can be regulated.

Several studies have confirmed the relevance of olfactory behavior with varying effects on interneuron survival. Two groups demonstrated that the survival of adult-born dopaminergic interneurons in the OB are activity dependent by closing, and then later opening, one naris (Bastien-Dionne et al., 2010; Sawada et al., 2011). A study on adult-born GABAergic interneurons in the OB found that olfactory enrichment only increased survival when it was present past the first 21 days from adult-born neuron birth, suggesting a critical period of integration that ensures survival (Bovetti et al., 2009). Learning is also important to survival but is dependent on the learning paradigm; only operant learning showed an effect on survival, while nonoperant learning did not (Mandairon et al., 2011).

Some of the studies that focused on a particular protein or growth factor correlated their findings with olfactory behavior. For example, connective tissue growth factor (CTGF), expressed in the granule cell layer, was found to decrease survival by enhancing the proapoptotic activity of the astrocyte-released growth factor, TFG-β2 (Khodosevich et al., 2013). When expression of CTGF was prevented, they observed increased spontaneous IPSC in the adult-born PG cells, which had a subsequent effect on mitral cell physiology. These changes in neural circuitry resulted in enhanced odor discrimination between two similar odorants (Khodosevich et al., 2013). This study demonstrates how changes at the molecular level affect organismal behavior.

Upon arrival in the OB, neuroblasts switch from tangential to radial migration. Radial migration is required in order for differentiation to commence. The first step is for single neuroblasts to detach from neuroblast chains. PK2 and a secreted ECM molecule called Reelin are important for detaching single cells from the neuroblast chains (Hack et al., 2002; Ng, 2005). Tenascin-R is an ECM molecule with several functional domains that underlie its many roles. In the OB, Tenascin-R also contributes to the detachment of neuroblasts from chains, and then directs neuroblasts to activity-dependent targets (Saghatelyan et al., 2004). Indeed, grafting Tenascin-R to a region outside the OB directed neuroblasts to the new location.

Maturation and Integration

In order for adult-born neurons to integrate into the existing OB circuitry, dendritic formation is necessary. Integration is activity dependent, both electrically and behaviorally. For example, interaction with one's young increases dendritic length and branch nodes in adult-born interneurons in the sheep OB (Brus et al., 2010). However, much more information is known about the electrophysiological integration of neuroblasts, specifically the role of GABA. The neurotransmitter GABA acts in a number of ways to ensure the integration of adult-born neurons. One role for GABA is to stimulate the dendritic growth of adult-born neurons integrating in the OB (Gascon et al., 2006). Similar to immature neurons in the dentate gyrus, activation of $GABA_A$ receptors is initially depolarizing and only during this period does GABA affect dendritic development (Gascon et al., 2006). GABA also promotes *lamellipodia* formation and has a stabilizing effect on newly formed dendrites through the polymerization of microtubules, which can be blocked by bicuculline, a GABA receptor antagonist.

Immature, adult-born granule cells in the OB receive synaptic input on their cell bodies before their dendritic arbors integrate into the existing circuitry in the EPL (Whitman and Greer, 2007b). One study demonstrated that network activity, likely from mature interneurons, provides the GABA input that supports dendritic development and synaptic integration (Akiba et al., 2009). In dopaminergic neurons, TH expression occurs after dendritic development is complete and is a final marker of maturation (Akiba et al., 2009). While depolarizing activity is important for TH expression, the presence of GABA further upregulates TH expression in new adult-born OB interneurons (Akiba et al., 2009). Thus GABA activity is crucial for the integration and maturation of interneurons.

RELEVANCE OF ADULT NEUROGENESIS TO OLFACTION

Mammals use the olfactory system for food detection and selection, social interactions, prey-predator interactions and the discernment of other types of danger. The olfactory system must respond to a wide range of odorants, and encode them in a context-dependent manner for subsequent, relevant retrieval. Olfaction is influenced by behavioral states and can dictate a behavioral response (compare food purchases during states of satiety vs. hunger!). Synaptic inhibition by OB interneurons is necessary for odor discrimination and perceptual learning (Abraham et al., 2010; Mandairon et al., 2008). The inhibitory effect of OB interneurons refines mitral/tufted cell activity creating a sparse network response that improves pattern separation (Lazarini and Lledo, 2011).

Studies have shown that there is some turnover of OB interneurons in the adult brain (Sakamoto et al., 2014; Mouret et al., 2009). Adult neurogenesis supplies a constant replacement of new interneurons through life. However, when neurogenesis begins to decrease with age, there is a subsequent increase in the life of the existing neurons (Mobley et al., 2013; Sui et al., 2012). Thus circuit maintenance is likely not the only purpose for new neurons in the OB.

The role of adult neurogenesis on olfactory perception has been controversial. In one study an X-ray lesion to the SVZ found an effect on olfactory fear conditioning, while a study that used an anti-mitotic infusion to the SVZ for 28 days found no effect on odor discrimination or long-term odor-associative memory tasks (Valley et al., 2009; Breton-Provencher et al., 2009). Conversely, in the FMRP knockout mice there is a significant increase in the number of immature dendritic spines that integrate into the OB circuitry. These mice have a deficit in odor sensitivity, but not in odor habituation or discrimination (Schilit-Nitenson). Furthermore, when a pan-caspase inhibitor was used to block apoptosis in the OB, there was a deficit in olfactory discrimination (Mouret et al., 2009). These studies suggest that a balance in turnover and integration of new neurons is important for olfaction.

Interactions with cortical input must be taken into account when considering the role of new neurons. The behavior state of OB processing is modified by centrifugal inputs from higher brain regions that mainly synapse on granule cells and provide context to the olfactory input (Lazarini and Lledo, 2011; Macrides et al., 1981). Changes in adult neurogenesis have significant effects on olfactory behavior when higher cognitive areas of the brain are involved. For example, decreased adult neurogenesis compromises long-term olfactory memory, olfactory fear conditioning and perceptual learning (Lazarini et al., 2009; Moreno et al., 2009; Sultan, 2010; Valley et al., 2009). Exposing mice to an odor-enriched environment during the critical survival period increased long-term odor memory (Rochefort et al., 2002; Rochefort and Lledo, 2005). At the level of the synapse, immature granule cells in the OB display long-term potentiation (LTP) when stimulated at the apical dendrite near the cell body, consistent with the location of the first synaptic input (Nissant et al., 2009; Whitman and Greer, 2009). These studies demonstrate the importance of adult neurogenesis to olfactory learning and memory.

SUMMARY

The production of new neurons in the SVZ occurs throughout life and contributes to behaviors important for navigating and adapting to a constantly changing olfactory environment. Though progress has been made, there is still much we do not understand regarding the regulation and behaviors of new cells that are produced. Specifically, the later stages of differentiation, integration and maturation require more research in order to determine how they are regulated. The clinical relevance of adult neurogenesis cannot be overstated. Stroke and traumatic brain injury outcomes as well as neurodegenerative diseases could be addressed with an influx of new neurons produced by the patient if they could be reprogrammed to target the correct brain region, adopt an appropriate fate and integrate into the existing circuitry. One method that brings us closer to achieving this goal is using reprogrammed iPSCs that are transplanted into the damaged area. This approach has been tested extensively using various types of stem cells on animal models of stroke (Lindvall and Kokaia, 2011). Some clinical trials have been carried out, or are underway, providing some evidence for safety and efficacy (Moniche et al., 2015; San Roman et al., 2015; Steinberg et al., 2016; Taguchi et al., 2015). Knowledge gained from studying adult neurogenesis can be applied to clinical studies being done to improve patient outcomes.

Although there are differences between adult neurogenesis in the forebrain and the dentate gyrus, there are many similarities. At first glance they are anatomically very different; however, aside from the ependymal cells of the lateral ventricles, the proliferative niches contain the same cell types. The behaviors of the proliferating, migrating and differentiating cells are similar as well. What is not clear is whether the mechanisms used by specific cell types for each behavior are the same between the two regions. Chapters 6 and 7 provide an outline for further investigation to compare the behaviors by the cells in these regions and the underlying mechanisms. Answers to questions in the SVZ may be found in the dentate gyrus and then tested in the SVZ and vice versa. It could be a systematic way to make discoveries in both proliferative areas.

DEFINITIONS

Centrifugal Axon projections from higher, more central processing areas of the brain.

Chemoattraction Movement of a cell toward a detected substance or molecule.

Chemorepulsion Movement of a cell away from a detected substance or molecule.

Dendrodentritic Synapses that form between neuronal dendrites.

Glomerulus A cluster of axonal synaptic terminals and dendritic endings in the OB.

Glycoproteins A class of proteins that has carbohydrate groups, called oligosaccharide chains, attached to amino acid side chains.

Lamellipodia Latin for "thin sheet" and "foot." Webbing-like sheets of actin that extend forward or retract like a wave; a morphological property of migrating cells.

Lateral inhibition The phenomenon that occurs when a neuron undergoes an action potential and releases inhibitory neurotransmitter onto neighboring cells.

Motor protein A class of proteins that move along the cytoskeleton by using ATP energy that causes a conformational change that translates into movement.

Tyrosine hydroxylase (TH) The molecule that is the precursor to dopamine. Often used as a marker of dopaminergic neurons in histology studies.

References

Abraham, N.M., Egger, V., Shimshek, D.R., Renden, R., Fukunaga, I., Sprengel, R., Seeburg, P.H., Klugmann, M., Margrie, T.W., Schaefer, A.T., Kuner, T., 2010. Synaptic inhibition in the olfactory bulb accelerates odor discrimination in mice. Neuron 65 (3), 399–411.

Aguirre, A., Rubio, M.E., Gallo, V., 2010. Notch and EGFR pathway interaction regulates neural stem cell number and self-renewal. Nature 467, 323–327.

Akiba, Y., Sasaki, H., Huerta, P.T., Estevez, A.G., Baker, H., Cave, J.W., 2009. γ-Aminobutyric acid-mediated regulation of the activity-dependent olfactory bulb dopaminergic phenotype. J. Neurosci. Res. 87, 2211–2221.

Alfonso, J., Penkert, H., Duman, C., Zuccotti, A., Monyer, H., 2015. Downregulation of sphingosine 1-phosphate receptor 1 promotes the switch from tangential to radial migration in the OB. J. Neurosci. 35, 13659–13672.

Alvarez-Buylla, A., Garcia-Verdugo, J.M., 2002. Neurogenesis in adult subventricular zone. J. Neurosci. 22, 629–634.

Andreu-Agulló, C., Morante-Redolat, J.M., Delgado, A.C., Fariñas, I., 2009. Vascular niche factor PEDF modulates Notch-dependent stemness in the adult subependymal zone. Nat. Neurosci. 12, 1514–1523.

Bardy, C., Alonso, M., Bouthour, W., Lledo, P.-M., 2010. How, when, and where new inhibitory neurons release neurotransmitters in the adult olfactory bulb. J. Neurosci. 30, 17023–17034.

Bartel, D.L., Rela, L., Hsieh, L., Greer, C.A., 2015. Dendrodendritic synapses in the mouse olfactory bulb external plexiform layer: dendrodendritic synapses of mouse EPL. J. Comp. Neurol. 523, 1145–1161.

Bastien-Dionne, P.-O., David, L.S., Parent, A., Saghatelyan, A., 2010. Role of sensory activity on chemospecific populations of interneurons in the adult olfactory bulb. J. Comp. Neurol. 518, 1847–1861.

Battista, D., Rutishauser, U., 2010. Removal of polysialic acid triggers dispersion of subventricularly derived neuroblasts into surrounding CNS tissues. J. Neurosci. 30, 3995–4003.

Berg, D.A., Belnoue, L., Song, H., Simon, A., 2013. Neurotransmitter-mediated control of neurogenesis in the adult vertebrate brain. Development 140, 2548–2561.

Bolteus, A.J., Bordey, A., 2004. GABA release and uptake regulate neuronal precursor migration in the postnatal subventricular zone. J. Neurosci. 24, 7623–7631.

Bovetti, S., Bovolin, P., Perroteau, I., Puche, A.C., 2007. Subventricular zone-derived neuroblast migration to the olfactory bulb is modulated by matrix remodelling. Eur. J. Neurosci. 25, 2021–2033.

Bovetti, S., Veyrac, A., Peretto, P., Fasolo, A., De Marchis, S., 2009. Olfactory enrichment influences adult neurogenesis modulating GAD67 and plasticity-related molecules expression in newborn cells of the olfactory bulb. PLoS One 4, e6359.

Bozoyan, L., Khlghatyan, J., Saghatelyan, A., 2012. Astrocytes control the development of the migration-promoting vasculature scaffold in the postnatal brain via VEGF signaling. J. Neurosci. 32, 1687–1704.

Breton-Provencher, V., Lemasson, M., Peralta III, M.R., Saghatelyan, A., 2009. Interneurons produced in adulthood are required for the normal functioning of the olfactory bulb network and for the execution of selected olfactory behaviors. J. Neurosci. 29 (48), 15245–15257.

Brus, M., Meurisse, M., Franceschini, I., Keller, M., Lévy, F., 2010. Evidence for cell proliferation in the sheep brain and its down-regulation by parturition and interactions with the young. Horm. Behav. 58, 737–746.

Bunk, E.C., Ertaylan, G., Ortega, F., Pavlou, M.A., Gonzalez Cano, L., Stergiopoulos, A., Safaiyan, S., Völs, S., van Cann, M., Politis, P.K., et al., 2016. Prox1 is required for oligodendrocyte cell identity in adult neural stem cells of the subventricular zone: oligodendrocyte cell identity in adult neural stem cells. Stem Cells 34, 2115–2129.

Carleton, A., Petreanu, L.T., Lansford, R., Alvarez-Buylla, A., Lledo, P.-M., 2003. Becoming a new neuron in the adult olfactory bulb. Nat. Neurosci. 6, 507–518.

Chazal, G., Durbec, P., Jankovski, A., Rougon, G., Cremer, H., 2000. Consequences of neural cell adhesion molecule deficiency on cell migration in the rostral migratory stream of the mouse. J. Neurosci. 20 (4), 1446–1457.
Chenn, A., Walsh, C.A., 2002. Regulation of cerebral cortical size by control of cell cycle exit in neural precursors. Science 297 (5580), 365–369.
Chiaramello, S., Dalmasso, G., Bezin, L., Marcel, D., Jourdan, F., Peretto, P., Fasolo, A., De Marchis, S., 2007. BDNF/TrkB interaction regulates migration of SVZ precursor cells via PI3-K and MAP-K signalling pathways: BDNF regulates SVZ neuroblast migration. Eur. J. Neurosci. 26, 1780–1790.
Colak, D., Mori, T., Brill, M.S., Pfeifer, A., Falk, S., Deng, C., Monteiro, R., Mummery, C., Sommer, L., Gotz, M., 2008. Adult neurogenesis requires Smad4-mediated bone morphogenic protein signaling in stem cells. J. Neurosci. 28, 434–446.
Cooper, J.A., 2013. Mechanisms of cell migration in the nervous system. J. Cell Biol. 202, 725–734.
Daynac, M., Tirou, L., Faure, H., Mouthon, M.-A., Gauthier, L.R., Hahn, H., Boussin, F.D., Ruat, M., 2016. Hedgehog controls quiescence and activation of neural stem cells in the adult ventricular-subventricular zone. Stem Cell Rep. 7, 735–748.
Díaz, D., Muñoz-Castañeda, R., Ávila-Zarza, C., Carretero, J., Alonso, J.R., Weruaga, E., 2017. Olfactory bulb plasticity ensures proper olfaction after severe impairment in postnatal neurogenesis. Sci. Rep. 7, .
Doetsch, F., García-Verdugo, J.M., Alvarez-Buylla, A., 1997. Cellular composition and three-dimensional organization of the subventricular germinal zone in the adult mammalian brain. J. Neurosci. 17, 5046–5061.
Doetsch, F., Caillé, I., Lim, D.A., García-Verdugo, J.M., Alvarez-Buylla, A., 1999. Subventricular zone astrocytes are neural stem cells in the adult mammalian brain. Cell 97, 703–716.
Emsley, J., 2003. α6β1 integrin directs migration of neuronal precursors in adult mouse forebrain. Exp. Neurol. 183, 273–285.
Ericson, J., Morton, S., Kawakami, A., Roelink, H., Jessell, T.M., 1996. Two critical periods of sonic hedgehog signaling required for the specification of motor neuron identity. Cell 87, 661–673.
Ferrón, S.R., Charalambous, M., Radford, E., McEwen, K., Wildner, H., Hind, E., Morante-Redolat, J.M., Laborda, J., Guillemot, F., Bauer, S.R., et al., 2011. Postnatal loss of Dlk1 imprinting in stem cells and niche astrocytes regulates neurogenesis. Nature 475, 381–385.
Gascon, E., Dayer, A.G., Sauvain, M.O., Potter, G., Jenny, B., De Roo, M., Zgraggen, E., Demaurex, N., Muller, D., Kiss, J.Z., 2006. GABA regulates dendritic growth by stabilizing lamellipodia in newly generated interneurons of the olfactory bulb. J. Neurosci. 26 (50), 12956–12966.
Girard, F., Eichenberger, S., Celio, M.R., 2014. Thrombospondin 4 deficiency in mouse impairs neuronal migration in the early postnatal and adult brain. Mol. Cell. Neurosci. 61, 176–186.
Gotts, J.E., Chesselet, M.-F., 2005. Mechanisms of subventricular zone expansion after focal cortical ischemic injury. J. Comp. Neurol. 488, 201–214.
Gritti, A., Frölichsthal-Schoeller, P., Galli, R., Parati, E.A., Cova, L., Pagano, S.F., Bjornson, C.R., Vescovi, A.L., 1999. Epidermal and fibroblast growth factors behave as mitogenic regulators for a single multipotent stem cell-like population from the subventricular region of the adult mouse forebrain. J. Neurosci. 19, 3287–3297.
Hack, I., Bancila, M., Loulier, K., Carroll, P., Cremer, H., 2002. Reelin is a detachment signal in tangential chain-migration during postnatal neurogenesis. Nat. Neurosci. 5, 939–945.
Herold, S., Jagasia, R., Merz, K., Wassmer, K., Lie, D.C., 2011. CREB signalling regulates early survival, neuronal gene expression and morphological development in adult subventricular zone neurogenesis. Mol. Cell. Neurosci. 46, 79–88.
High, F.A., Lu, M.M., Pear, W.S., Loomes, K.M., Kaestner, K.H., Epstein, J.A., 2008. Endothelial expression of the Notch ligand Jagged1 is required for vascular smooth muscle development. Proc. Natl. Acad. Sci. 105, 1955–1959.
Hinds, J.W., Hinds, P.L., 1976. Synapse formation in the mouse olfactory bulb. I. Quantitative studies. J. Comp. Neurol. 169, 15–40.

Ihrie, R.A., Shah, J.K., Harwell, C.C., Levine, J.H., Guinto, C.D., Lezameta, M., Kriegstein, A.R., Alvarez-Buylla, A., 2011. Persistent sonic hedgehog signaling in adult brain determines neural stem cell positional identity. Neuron 71, 250–262.

Imayoshi, I., Kageyama, R., 2011. The role of notch signaling in adult neurogenesis. Mol. Neurobiol. 44, 7–12.

Jacques, T.S., Relvas, J.B., Nishimura, S., Pytela, R., Edwards, G.M., Streuli, C.H., ffrench-Constant, C., 1998. Neural precursor cell chain migration and division are regulated through different beta1 integrins. Development 125, 3167–3177.

Karalay, O., Doberauer, K., Vadodaria, K.C., Knobloch, M., Berti, L., Miquelajauregui, A., Schwark, M., Jagasia, R., Taketo, M.M., Tarabykin, V., et al., 2011. Prospero-related homeobox 1 gene (Prox1) is regulated by canonical Wnt signaling and has a stage-specific role in adult hippocampal neurogenesis. Proc. Natl. Acad. Sci. 108, 5807–5812.

Karperien, A., Ahammer, H., Jelinek, H.F., 2013. Quantitating the subtleties of microglial morphology with fractal analysis. Front. Cell. Neurosci. 7, .

Kelsch, W., Lin, C.-W., Lois, C., 2008. Sequential development of synapses in dendritic domains during adult neurogenesis. Proc. Natl. Acad. Sci. 105, 16803–16808.

Kelsch, W., Lin, C.-W., Mosley, C.P., Lois, C., 2009. A critical period for activity-dependent synaptic development during olfactory bulb adult neurogenesis. J. Neurosci. 29, 11852–11858.

Kelsch, W., Li, Z., Eliava, M., Goengrich, C., Monyer, H., 2012. GluN2B-containing NMDA receptors promote wiring of adult-born neurons into olfactory bulb circuits. J. Neurosci. 32, 12603–12611.

Khodosevich, K., Lazarini, F., von Engelhardt, J., Kaneko, H., Lledo, P.-M., Monyer, H., 2013. Connective tissue growth factor regulates interneuron survival and information processing in the olfactory bulb. Neuron 79, 1136–1151.

Kirschenbaum, B., Doetsch, F., Lois, C., Alvarez-Buylla, A., 1999. Adult subventricular zone neuronal precursors continue to proliferate and migrate in the absence of the olfactory bulb. J. Neurosci. 19, 2171–2180.

Kohwi, M., Petryniak, M.A., Long, J.E., Ekker, M., Obata, K., Yanagawa, Y., Rubenstein, J.L.R., Alvarez-Buylla, A., 2007. A subpopulation of olfactory bulb GABAergic interneurons is derived from Emx1- and Dlx5/6-expressing progenitors. J. Neurosci. 27, 6878–6891.

Kosaka, K., Kosaka, T., 2007. Chemical properties of type 1 and type 2 periglomerular cells in the mouse olfactory bulb are different from those in the rat olfactory bulb. Brain Res. 1167, 42–55.

Kosodo, Y., Röper, K., Haubensak, W., Marzesco, A.M., Corbeil, D., Huttner, W.B., 2004. Asymmetric distribution of the apical plasma membrane during neurogenic divisions of mammalian neuroepithelial cells. EMBO J. 23 (11), 2314–2324.

Lazarini, F., Lledo, P.-M., 2011. Is adult neurogenesis essential for olfaction? Trends Neurosci. 34, 20–30.

Lazarini, F., Mouthon, M.A., Gheusi, G., de Chaumont, F., Olivo-Marin, J.C., Lamarque, S., Abrous, D.N., Boussin, F.D., Lledo, P.M., 2009. Cellular and behavioral effects of cranial irradiation of the subventricular zone in adult mice. PLoS ONE 4 (9), e7017.

Lehtinen, M.K., Zappaterra, M.W., Chen, X., Yang, Y.J., Hill, A.D., Lun, M., Maynard, T., Gonzalez, D., Kim, S., Ye, P., et al., 2011. The cerebrospinal fluid provides a proliferative niche for neural progenitor cells. Neuron 69, 893–905.

Lemasson, M., 2005. Neonatal and adult neurogenesis provide two distinct populations of newborn neurons to the mouse olfactory bulb. J. Neurosci. 25, 6816–6825.

Lepousez, G., Valley, M.T., Lledo, P.M., 2013. The impact of adult neurogenesis on olfactory bulb circuits and computations. Annu. Rev. Physiol. 75, 339–363.

Li, T., Pan, Y.-W., Wang, W., Abel, G., Zou, J., Xu, L., Storm, D.R., Xia, Z., 2013. Targeted deletion of the ERK5 MAP kinase impairs neuronal differentiation, migration, and survival during adult neurogenesis in the olfactory bulb. PLoS ONE 8, e61948.

Lim, D.A., Tramontin, A.D., Trevejo, J.M., Herrera, D.G., García-Verdugo, J.M., Alvarez-Buylla, A., 2000. Noggin antagonizes BMP signaling to create a niche for adult neurogenesis. Neuron 28, 713–726.

Lin, C.-W., Sim, S., Ainsworth, A., Okada, M., Kelsch, W., Lois, C., 2010. Genetically increased cell-intrinsic excitability enhances neuronal integration into adult brain circuits. Neuron 65, 32–39.

Lindvall, O., Kokaia, Z., 2011. Stem cell research in stroke: how far from the clinic? Stroke 42, 2369–2375.

Litingtung, Y., Chiang, C., 2000. Control of Shh activity and signaling in the neural tube. Dev. Dyn. 219, 143–154.

Liu, X., Wang, Q., Haydar, T.F., Bordey, A., 2005. Nonsynaptic GABA signaling in postnatal subventricular zone controls proliferation of GFAP-expressing progenitors. Nat. Neurosci. 8, 1179–1187.

Lledo, P.-M., Saghatelyan, A., 2005. Integrating new neurons into the adult olfactory bulb: joining the network, life-death decisions, and the effects of sensory experience. Trends Neurosci. 28, 248–254.

Lledo, P.-M., Merkle, F.T., Alvarez-Buylla, A., 2008. Origin and function of olfactory bulb interneuron diversity. Trends Neurosci. 31, 392–400.

Lois, C., García-Verdugo, J.M., Alvarez-Buylla, A., 1996. Chain migration of neuronal precursors. Science 271, 978–981.

Luo, J., Chen, X., Pan, Y.-W., Lu, S., Xia, Z., Storm, D.R., 2015. The type 3 adenylyl cyclase is required for the survival and maturation of newly generated granule cells in the olfactory bulb. PLoS ONE 10, e0122057.

Luskin, M.B., Boone, M.S., 1994. Rate and pattern of migration of lineally-related olfactory bulb interneurons generated postnatally in the subventricular zone of the rat. Chem. Senses 19, 695–714.

Macrides, F., Davis, B.J., Youngs, W.M., Nadi, N.S., Margolis, F.L., 1981. Cholinergic and catecholaminergic afferents to the olfactory bulb in the hamster: a neuroanatomical, biochemical, and histochemical investigation. J. Comp. Neurol. 203, 495–514.

Malnic, B., Hirono, J., Sato, T., Buck, L.B., 1999. Combinatorial receptor codes for odors. Cell 96, 713–723.

Mandairon, N., Didier, A., Linster, C., 2008. Odor enrichment increases interneurons responsiveness in spatially defined regions of the olfactory bulb correlated with perception. Neurobiol. Learn. Mem. 90 (1), 178–184.

Mandairon, N., Sultan, S., Nouvian, M., Sacquet, J., Didier, A., 2011. Involvement of newborn neurons in olfactory associative learning? The operant or non-operant component of the task makes all the difference. J. Neurosci. 31, 12455–12460.

Mathieu, C., Sii-Felice, K., Fouchet, P., Etienne, O., Haton, C., Mabondzo, A., Boussin, F.D., Mouthon, M.-A., 2008. Endothelial cell-derived bone morphogenetic proteins control proliferation of neural stem/progenitor cells. Mol. Cell. Neurosci. 38, 569–577.

Menn, B., Garcia-Verdugo, J.M., Yaschine, C., Gonzalez-Perez, O., Rowitch, D., Alvarez-Buylla, A., 2006. Origin of oligodendrocytes in the subventricular zone of the adult brain. J. Neurosci. 26, 7907–7918.

Merkle, F.T., Mirzadeh, Z., Alvarez-Buylla, A., 2007. Mosaic organization of neural stem cells in the adult brain. Science 317, 381–384.

Mirzadeh, Z., Merkle, F.T., Soriano-Navarro, M., Garcia-Verdugo, J.M., Alvarez-Buylla, A., 2008. Neural stem cells confer unique pinwheel architecture to the ventricular surface in neurogenic regions of the adult brain. Cell Stem Cell 3, 265–278.

Mitro, A., Palkovits, M., 1981. Morphology of the rat brain ventricles, ependyma, and periventricular structures. Bibl. Anat. 1–110.

Mobley, A.S., Bryant, A.K., Richard, M.B., Brann, J.H., Firestein, S.J., Greer, C.A., 2013. Age-dependent regional changes in the rostral migratory stream. Neurobiol. Aging 34, 1873–1881.

Moniche, F., Escudero, I., Zapata-Arriaza, E., Usero-Ruiz, M., Prieto-León, M., de la Torre, J., Gamero, M.-A., Tamayo, J.A., Ochoa-Sepúlveda, J.-J., Maestre, J., et al., 2015. Intra-arterial bone marrow mononuclear cells (BM-MNCS) transplantation in acute ischemic stroke (IBiS trial): protocol of a phase II, randomized, dose-finding, controlled multicenter trial. Int. J. Stroke 10, 1149–1152.

Morell, M., Tsan, Y., O'Shea, K.S., 2015. Inducible expression of noggin selectively expands neural progenitors in the adult SVZ. Stem Cell Res. 14, 79–94.

Moreno, M.M., Linster, C., Escanilla, O., Sacquet, J., Didier, A., Mandairon, N., 2009. Olfactory perceptual learning requires adult neurogenesis. Proc Natl Acad Sci USA 106 (42), 17980–17985.

Mouret, A., Lepousez, G., Gras, J., Gabellec, M.M., Lledo, P.M., 2009. Turnover of newborn olfactory bulb neurons optimizes olfaction. J Neurosci. 29 (39), 12302–12314.

Murase, S., Horwitz, A.F., 2002. Deleted in colorectal carcinoma and differentially expressed integrins mediate the directional migration of neural precursors in the rostral migratory stream. J. Neurosci. 22, 3568–3579.

Nagayama, S., Homma, R., Imamura, F., 2014. Neuronal organization of olfactory bulb circuits. Front. Neural Circ. 8, .

Najac, M., Sanz Diez, A., Kumar, A., Benito, N., Charpak, S., De Saint Jan, D., 2015. Intraglomerular lateral inhibition promotes spike timing variability in principal neurons of the olfactory bulb. J. Neurosci. 35, 4319–4331.

Nam, H., Benezra, R., 2009. High levels of Id1 expression define B1 type adult neural stem cells. Cell Stem Cell 5, 515–526.

Ng, K.L., 2005. Dependence of olfactory bulb neurogenesis on prokineticin 2 signaling. Science 308, 1923–1927.

Nguyen-Ba-Charvet, K.T., Picard-Riera, N., Tessier-Lavigne, M., Baron-Van Evercooren, A., Sotelo, C., Chédotal, A., 2004. Multiple roles for slits in the control of cell migration in the rostral migratory stream. J. Neurosci. 24, 1497–1506.

Nimmerjahn, A., 2005. Resting microglial cells are highly dynamic surveillants of brain parenchyma in vivo. Science 308, 1314–1318.

Ninkovic, J., Pinto, L., Petricca, S., Lepier, A., Sun, J., Rieger, M.A., Schroeder, T., Cvekl, A., Favor, J., Götz, M., 2010. The transcription factor Pax6 regulates survival of dopaminergic olfactory bulb neurons via Crystallin αA. Neuron 68, 682–694.

Nissant, A., Bardy, C., Katagiri, H., Murray, K., Lledo, P.M., 2009. Adult neurogenesis promotes synaptic plasticity in the olfactory bulb. Nat. Neurosci. 12 (6), 728–730.

Ottone, C., Parrinello, S., 2015. Multifaceted control of adult SVZ neurogenesis by the vascular niche. Cell Cycle 14, 2222–2225.

Paez-Gonzalez, P., Asrican, B., Rodriguez, E., Kuo, C.T., 2014. Identification of distinct ChAT+ neurons and activity-dependent control of postnatal SVZ neurogenesis. Nat. Neurosci. 17, 934–942.

Parent, J.M., Vexler, Z.S., Gong, C., Derugin, N., Ferriero, D.M., 2002. Rat forebrain neurogenesis and striatal neuron replacement after focal stroke. Ann. Neurol. 52, 802–813.

Pastrana, E., Cheng, L.-C., Doetsch, F., 2009. Simultaneous prospective purification of adult subventricular zone neural stem cells and their progeny. Proc. Natl. Acad. Sci. U. S. A. 106, 6387–6392.

Peretto, P., Dati, C., De Marchis, S., Kim, H.H., Ukhanova, M., Fasolo, A., Margolis, F.L., 2004. Expression of the secreted factors noggin and bone morphogenetic proteins in the subependymal layer and olfactory bulb of the adult mouse brain. Neuroscience 128, 685–696.

Petreanu, L., Alvarez-Buylla, A., 2002. Maturation and death of adult-born olfactory bulb granule neurons: role of olfaction. J. Neurosci. 22, 6106–6113.

Ponti, G., Obernier, K., Guinto, C., Jose, L., Bonfanti, L., Alvarez-Buylla, A., 2013. Cell cycle and lineage progression of neural progenitors in the ventricular-subventricular zones of adult mice. Proc. Natl. Acad. Sci. 110, E1045–E1054.

Ribeiro Xavier, A.L., Kress, B.T., Goldman, S.A., Lacerda de Menezes, J.R., Nedergaard, M., 2015. A distinct population of microglia supports adult neurogenesis in the subventricular zone. J. Neurosci. 35, 11848–11861.

Rochefort, C., Lledo, P.M., 2005. Short-term survival of newborn neurons in the adult olfactory bulb after exposure to a complex odor environment. Eur. J. Neurosci. 22 (11), 2863–2870.

Rochefort, C., Gheusi, G., Vincent, J.D., Lledo, P.M., 2002. Enriched odor exposure increases the number of newborn neurons in the adult olfactory bulb and improves odor memory. J. Neurosci. 22 (7), 2679–2689.

Rousselot, P., Lois, C., Alvarez-Buylla, A., 1995. Embryonic (PSA) N-CAM reveals chains of migrationg neuroblasts between the lateral ventricle and the olfactory bulb of adult mice. J. Comp. Neurol. 351, 51–61.

Roybon, L., Deierborg, T., Brundin, P., Li, J.-Y., 2009. Involvement of Ngn2, Tbr and NeuroD proteins during postnatal olfactory bulb neurogenesis. Eur. J. Neurosci. 29, 232–243.

Rutishauser, U., Watanabe, M., Silver, J., Troy, F.A., Vimr, E.R., 1985. Specific alteration of NCAM-mediated cell adhesion by an endoneuraminidase. J. Cell Biol. 101, 1842–1849.

Saghatelyan, A., de Chevigny, A., Schachner, M., Lledo, P.-M., 2004. Tenascin-R mediates activity-dependent recruitment of neuroblasts in the adult mouse forebrain. Nat. Neurosci. 7, 347–356.

Saghatelyan, A., Roux, P., Migliore, M., Rochefort, C., Desmaisons, D., Charneau, P., Shepherd, G.M., Lledo, P.-M., 2005. Activity-dependent adjustments of the inhibitory network in the olfactory bulb following early postnatal deprivation. Neuron 46, 103–116.

Saha, B., Ypsilanti, A.R., Boutin, C., Cremer, H., Chedotal, A., 2012. Plexin-B2 regulates the proliferation and migration of neuroblasts in the postnatal and adult subventricular zone. J. Neurosci. 32, 16892–16905.

Sakamoto, M., Kageyama, R., Imayoshi, I., 2014. The functional significance of newly born neurons integrated into olfactory bulb circuits. Front. Neurosci. 8, 121.

San Roman, J.A., Sánchez, P.L., Villa, A., Sanz-Ruiz, R., Fernandez-Santos, M.E., Gimeno, F., Ramos, B., Arnold, R., Serrador, A., Gutiérrez, H., et al., 2015. Comparison of different bone marrow–derived stem cell approaches in reperfused STEMI. J. Am. Coll. Cardiol. 65, 2372–2382.

Sawada, M., Kaneko, N., Inada, H., Wake, H., Kato, Y., Yanagawa, Y., Kobayashi, K., Nemoto, T., Nabekura, J., Sawamoto, K., 2011. Sensory input regulates spatial and subtype-specific patterns of neuronal turnover in the adult olfactory bulb. J. Neurosci. 31, 11587–11596.

Sawamoto, K., Wichterle, H., Gonzalez-Perez, O., Cholfin, J.A., Yamada, M., Spassky, N., Murcia, N.S., Garcia-Verdugo, J.M., Marin, O., Rubenstein, J.L.R., et al., 2006. New neurons follow the flow of cerebrospinal fluid in the adult brain. Science 311, 629–632.

Schaar, B.T., McConnell, S.K., 2005. Cytoskeletal coordination during neuronal migration. Proc. Natl. Acad. Sci. 102, 13652–13657.

Scotto-Lomassese, S., Nissant, A., Mota, T., Neant-Fery, M., Oostra, B.A., Greer, C.A., Lledo, P.-M., Trembleau, A., Caille, I., 2011. Fragile X mental retardation protein regulates new neuron differentiation in the adult olfactory bulb. J. Neurosci. 31, 2205–2215.

Shen, Q., 2004. Endothelial cells stimulate self-renewal and expand neurogenesis of neural stem cells. Science 304, 1338–1340.

Shen, Q., Wang, Y., Kokovay, E., Lin, G., Chuang, S.-M., Goderie, S.K., Roysam, B., Temple, S., 2008. Adult SVZ stem cells lie in a vascular niche: a quantitative analysis of niche cell-cell interactions. Cell Stem Cell 3, 289–300.

Snapyan, M., Lemasson, M., Brill, M.S., Blais, M., Massouh, M., Ninkovic, J., Gravel, C., Berthod, F., Götz, M., Barker, P.A., et al., 2009. Vasculature guides migrating neuronal precursors in the adult mammalian forebrain via brain-derived neurotrophic factor signaling. J. Neurosci. 29, 4172–4188.

Soria-Gómez, E., Bellocchio, L., Reguero, L., Lepousez, G., Martin, C., Bendahmane, M., Ruehle, S., Remmers, F., Desprez, T., Matias, I., et al., 2014. The endocannabinoid system controls food intake via olfactory processes. Nat. Neurosci. 17, 407–415.
Steinberg, G.K., Kondziolka, D., Wechsler, L.R., Lunsford, L.D., Coburn, M.L., Billigen, J.B., Kim, A.S., Johnson, J.N., Bates, D., King, B., et al., 2016. Clinical outcomes of transplanted modified bone marrow–derived mesenchymal stem cells in stroke: a phase 1/2a study. Stroke 47, 1817–1824.
Stewart, R.R., Hoge, G.J., Zigova, T., Luskin, M.B., 2002. Neural progenitor cells of the neonatal rat anterior subventricular zone express functional GABAA receptors. J. Neurobiol. 50, 305–322.
Sui, Y., Horne, M.K., Stanić, D., 2012. Reduced proliferation in the adult mouse subventricular zone increases survival of olfactory bulb interneurons. PLoS ONE 7 (2), e31549.
Sultan, S., Mandairon, N., Kermen, F., Garcia, S., Sacquet, J., Didier, A., 2010. Learning-dependent neurogenesis in the olfactory bulb determines long-term olfactory memory. FASEB J. 24 (7), 2355–2363.
Sun, D., Bullock, M.R., McGinn, M.J., Zhou, Z., Altememi, N., Hagood, S., Hamm, R., Colello, R.J., 2009. Basic fibroblast growth factor-enhanced neurogenesis contributes to cognitive recovery in rats following traumatic brain injury. Exp. Neurol. 216, 56–65.
Taguchi, A., Sakai, C., Soma, T., Kasahara, Y., Stern, D.M., Kajimoto, K., Ihara, M., Daimon, T., Yamahara, K., Doi, K., et al., 2015. Intravenous autologous bone marrow mononuclear cell transplantation for stroke: phase 1/2a clinical trial in a homogeneous group of stroke patients. Stem Cells Dev. 24, 2207–2218.
Takano, T., Tian, G.-F., Peng, W., Lou, N., Libionka, W., Han, X., Nedergaard, M., 2006. Astrocyte-mediated control of cerebral blood flow. Nat. Neurosci. 9, 260–267.
Tavazoie, M., Van der Veken, L., Silva-Vargas, V., Louissaint, M., Colonna, L., Zaidi, B., Garcia-Verdugo, J.M., Doetsch, F., 2008. A specialized vascular niche for adult neural stem cells. Cell Stem Cell 3, 279–288.
Temple, S., 2001. The development of neural stem cells. Nature 414, 112–117.
Tong, C.K., Chen, J., Cebrián-Silla, A., Mirzadeh, Z., Obernier, K., Guinto, C.D., Tecott, L.H., García-Verdugo, J.M., Kriegstein, A., Alvarez-Buylla, A., 2014. Axonal control of the adult neural stem cell niche. Cell Stem Cell 14, 500–511.
Valero, J., Weruaga, E., Murias, A.R., Recio, J.S., Curto, G.G., Gómez, C., Alonso, J.R., 2007. Changes in cell migration and survival in the olfactory bulb of thepcd/pcd mouse. Dev. Neurobiol. 67, 839–859.
Vallee, R.B., Seale, G.E., Tsai, J.-W., 2009. Emerging roles for myosin II and cytoplasmic dynein in migrating neurons and growth cones. Trends Cell Biol. 19, 347–355.
Valley, M.T., Mullen, T.R., Schultz, L.C., Sagdullaev, B.T., Firestein, S., 2009. Ablation of mouse adult neurogenesis alters olfactory bulb structure and olfactory fear conditioning. Front Neurosci. 3, 51.
Walton, N.M., Sutter, B.M., Laywell, E.D., Levkoff, L.H., Kearns, S.M., Marshall, G.P., Scheffler, B., Steindler, D.A., 2006. Microglia instruct subventricular zone neurogenesis. Glia 54, 815–825.
Wang, L., 2006. Matrix metalloproteinase 2 (MMP2) and MMP9 secreted by erythropoietin-activated endothelial cells promote neural progenitor cell migration. J. Neurosci. 26, 5996–6003.
Whitman, M.C., Greer, C.A., 2007a. Adult-generated neurons exhibit diverse developmental fates. Dev. Neurobiol. 67, 1079–1093.
Whitman, M.C., Greer, C.A., 2007b. Synaptic integration of adult-generated olfactory bulb granule cells: basal axodendritic centrifugal input precedes apical dendrodendritic local circuits. J. Neurosci. 27, 9951–9961.
Whitman, M.C., Greer, C.A., 2009. Adult neurogenesis and the olfactory system. Prog. Neurobiol. 89 (2), 162–175.
Whitman, M.C., Fan, W., Rela, L., Rodriguez-Gil, D.J., Greer, C.A., 2009. Blood vessels form a migratory scaffold in the rostral migratory stream. J. Comp. Neurol. 516, 94–104.

Wu, W., Wong, K., Chen, J., Jiang, Z., Dupuis, S., Wu, J.Y., Rao, Y., 1999. Directional guidance of neuronal migration in the olfactory system by the protein slit. Nature 400, 331–336.

Xiong, X.-Y., Liu, L., Yang, Q.-W., 2016. Functions and mechanisms of microglia/macrophages in neuroinflammation and neurogenesis after stroke. Prog. Neurobiol. 142, 23–44.

Yamadori, T., Nara, K., 1979. The directions of ciliary beat on the wall of the lateral ventricle and the currents of the cerebrospinal fluid in the brain ventricles. Scan. Electron Microsc. 335–340.

Yamaguchi, M., Mori, K., 2005. Critical period for sensory experience-dependent survival of newly generated granule cells in the adult mouse olfactory bulb. Proc. Natl. Acad. Sci. 102, 9697–9702.

Young, K.M., Fogarty, M., Kessaris, N., Richardson, W.D., 2007. Subventricular zone stem cells are heterogeneous with respect to their embryonic origins and neurogenic fates in the adult olfactory bulb. J. Neurosci. 27, 8286–8296.

CHAPTER

8

Adult Neurogenesis in Health and Disease

OUTLINE

LEARNING OBJECTIVES

1. Discuss how exercise and aging, independently or together affect adult neurogenesis.
2. Explain how the Morris Water Maze can be adapted to test spatial learning or stimulus-response learning.
3. List the connections between olfactory function, adult neurogenesis, aging and mortality.
4. Discuss the effects of antidepressant treatments on adult neurogenesis.

https://doi.org/10.1016/B978-0-12-811014-0.00008-1

5. Describe what is known about inflammation in epilepsy, stroke and degenerative diseases.
6. Describe the progress on stem cell replacement to treat stroke and degenerative diseases.
7. Discuss the challenges to manipulate adult neurogenesis or use stem cell replacement to treat spinal cord injury.

ADULT NEUROGENESIS IN HEALTH

Currently, worldwide life expectancy is 70.5 years of age. By 2050, life expectancy is expected to rise, and the percentage of the population older than 65 years is expected to jump from 8% to 17%. As these numbers increase, we expect to see a greater number of individuals suffering from noncommunicable and chronic diseases like atherosclerosis, cancer, diabetes, and dementia that all share increased life expectancy as a risk factor. Stem cell research has provided timely, important insights into treating or managing these types of diseases, if not necessarily a cure (Doeppner and Hermann, 2014; Voltarelli et al., 2007; Zhang et al., 2017). Early investigations on adult neurogenesis discovered its role in learning and memory, as well as the positive effects of novelty, enrichment and exercise. Now we will focus on (1) the effect of disease and disorders on adult neurogenesis and (2) how neurogenesis can be harnessed to prevent or mitigate the disease and disorders affecting the brain.

Environment

The first studies confirming adult neurogenesis in human and nonhuman primates caused a surge of interest in the regulation of adult neurogenesis in rodents. Earlier studies had shown that environmental enrichment had a profound effect on the hippocampus, including increased hippocampal thickness and complex dendritic arborization (Walsh et al., 1969; Fiala et al., 1978). Using BrdU or [3H]thymidine, they showed that environmental enrichment also increased the number of adult-born neurons in the dentate gyrus. Twenty-one-day old (weaned) mice were given either a standard home cage or an enriched cage that was much larger and included tunnels, toys, a running wheel and a nest, where they lived for 40 days. During the last 12 days they received daily injections of BrdU. The mice in the enriched cage differed significantly from the standard home cage mice in the number of BrdU-positive cells and in the decreased time it took to learn a spatial-navigation memory task called the Morris Water Maze (Box 8.1). The results provided early evidence that adult neurogenesis contributed to learning and memory. Additional experiments on

BOX 8.1

THE MORRIS WATER MAZE

The Morris Water Maze (MWM) consists of a large pool of opaque water surrounded by visual markers that can be easily observed by a mouse or rat placed in the water. The mouse swims around the pool until they find a platform hidden under the surface of the water. Through repetition, mice eventually learn the location of the platform using surrounding visual cues (D'Hooge and De Deyn, 2001; Morris, 1981). Varying the start position for the mouse requires spatial orientation and memory. By keeping a constant start position, the mouse learns the location of the platform through stimulus-response learning that does not require visual cues (Fig. 8.1).

In Chapter 6, we described the important role the hippocampus has in learning and memory, and briefly, how adult neurogenesis contributes to these processes. There is great interest in understanding the correlation between changes in adult neurogenesis and learning ability and memory performance. For example, participation in an associated learning (classical or operant conditioning) task doubled the number of new BrdU positive neurons produced in the dentate gyrus (Gould et al., 1999). Similarly, operant learning resulted in significant increases in both proliferation and survival of adult-born neurons in the OB. Thus the Morris Water Maze can be used to show improved learning with increased adult neurogenesis.

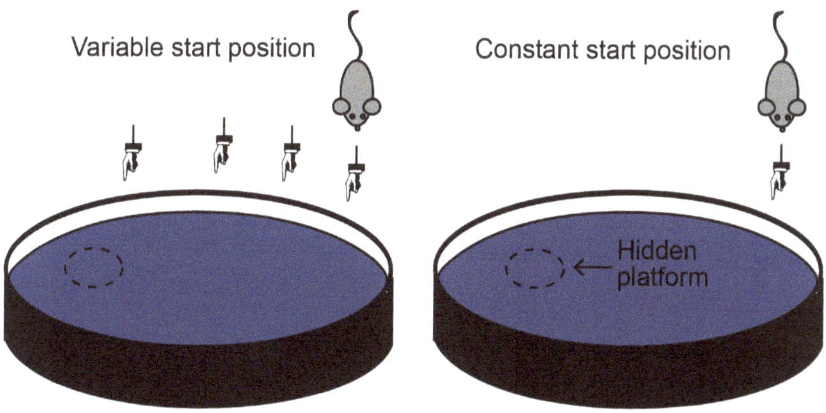

FIG. 8.1 The Morris Water Maze. A mouse is placed in a pool of opaque water that hides a platform just below the surface. When the start position is varied (left), the mouse learns the location of the platform in relation to cues surrounding the pool. The experiment tests spatial recognition and relational learning. When the start position is held constant (right), the experiment changes to a stimulus-response task—performing a particular behavior in response to a particular stimulus for a small food reward.

environmental enrichment that included a running wheel found that there was a significant increase in brain-derived neurotrophic factor (BDNF) that was also shown to enhance adult neurogenesis (Rossi et al., 2006). These experiments demonstrated a functional role for adult neurogenesis and the neuroprotective effect of environmental enrichment.

In order to understand which phase of adult neurogenesis was affected by environmental enrichment, mice received an injection of retrovirus that would label a small number of new adult-born neurons red. Mice were exposed to 48h of enrichment at different days following the injection, and evaluated 21 days postinjection (Alvarez et al., 2016). Based on observations of longer dendritic length, faster integration and enhanced connectivity, such as increased spine density, the critical period of enrichment was determined to be nine days after injection. Mature granule cells contribute to this phenomenon as well, by activating *parvalbumin*-positive interneurons, which then facilitate the accelerated integration of the adult-born cells (Alvarez et al., 2016). Thus exposure to environmental enrichment during the critical period favors survival and recruitment of adult-born cells.

Exercise

Inclusion of the running wheel in environmental enrichment experiments led some scientists to ask whether exercise alone affected adult neurogenesis. Further experiments were conducted that compared environmental enrichment with and without exercise. By comparing running (a cage with only running wheels) to environmental enrichment only, or environmental enrichment plus running, a study found that only running increased hippocampal proliferation and levels of BDNF (Kobilo et al., 2011). In middle-aged female mice, running improved memory function in addition to BDNF-enhanced adult neurogenesis in the hippocampus (Marlatt et al., 2012). Furthermore, running enhances learning, long-term potentiation (LTP) and spatial memory, linking its effects to hippocampus-dependent tasks (van Praag et al., 1999). The effect of exercise on adult neurogenesis appears to be limited to the dentate gyrus as there was no increase in neurogenesis or BDNF in the SVZ (Brown et al., 2003).

Exercise causes a number of positive effects on hippocampal neurons some of which are mediated by the release of a muscle secretory factor called Cathepsin B (CTSB) protein (Moon et al., 2016). Treadmill use by human and nonhuman primates increases plasma CTSB levels, and in mice, CTSB enhances spatial memory and adult neurogenesis in the dentate gyrus (Moon et al., 2016). Running causes changes to the innervation of dentate gyrus neurons, with increased afferents from sub-cortical, cortical regions, sensory cortices and regions important for spatial memory (Vivar et al., 2016). On adult-born cells, the effect of running causes increased total apical dendritic length that includes more branching (Sah et al., 2017). Similar to adult-born neurons in the OB, those in the dentate gyrus express GABA and NMDA receptors early in the differentiation process. Running increases NMDA receptor expression, which favors their survival (Carleton et al., 2003; Sah et al., 2017). However, the increased dendritic area requires a higher stimulation intensity to fully excite NMDA synapses. Thus more research is required to determine how these early changes in circuitry determine running-increased LTP and spatial memory. While adult neurogenesis per se is difficult to measure in humans, studies have correlated exercise with improved memory and affect (Déry et al., 2013; Stroth et al., 2009). As we discuss normal aging, disorders and disease, we will continue to refer to exercise.

Experience

Experience is difficult to define as an effector of adult neurogenesis. Novel experience may be a better term; however, learning and social interactions could all be considered novel experiences. Indeed, a large body of work has investigated the role of social experiences on adult neurogenesis including behaviors such as mating, reproduction and parenting (for review see Lieberwirth and Wang, 2012). These examples are all initially, novel experiences, and are associated with *chemosensory* cues.

Chemosensory cues provide a wealth of social information within a species (sexual/social stimuli). Male moths use single odorant compounds to detect the presence of a female, while in rodents, pheromones convey territory, estrous and suckling cues as well as other behaviors (Kaur et al., 2014; Muthukumar et al., 2013; Teicher et al., 1980; Zhang et al., 2015). Social stimuli increased proliferation and/or survival of adult-born neurons, with sexual experience affecting proliferation in the dentate gyrus and social stimuli affecting the SVZ (Glasper and Gould, 2013; Hawken et al., 2009; Smith et al., 2001). Conversely, negative or lack of social interactions (isolation) reduced adult neurogenesis, while increasing the incidents of anxiety and depression (Alpass and Neville, 2003; Gould et al., 1997; Lee and Robbins, 1998; Lieberwirth et al., 2012).

Experiments that measured adult neurogenesis during gestation have reported mixed data: decreased dentate gyrus neurogenesis in meadow voles, no effect in rats, and increased neurogenesis in mice depending on the day of gestation (Galea and McEwen, 1999). In the SVZ, the story is also unclear, as increases in proliferation in rats and mice were observed, depending on the day of gestation, but not in sheep (Brus et al., 2010; Furuta and Bridges, 2005; Shingo, 2003). Thus there appears to be species-specific, brain region-specific and temporal effects of gestation on adult neurogenesis.

The behaviors described previously involve hormones such as testosterone, estroge, and luteinizing hormone. In rats, estrogen increases proliferation in the dentate gyrus in a dose- and time-dependent manner. Although in one study a decrease in neuronal survival was observed (Barker and Galea, 2008; Tanapat et al., 1999). Testosterone also has a dose-dependent effect: social isolation (low levels) causes a decrease in adult neurogenesis, while administration of testosterone (high levels) in castrated male rats saw a further decrease in adult neurogenesis, but is hypothesized to buffer social isolation in uncastrated males (Spritzer et al., 2011). In contrast, testosterone increases survival of adult-born hippocampal neurons in castrated meadow voles (Ormerod et al., 2004). Luteinizing hormone, prolactin and oxytocin—hormones that regulate reproductive behaviors—all show an increasing effect on adult neurogenesis in the animal models and paradigms tested (Hawken et al., 2009; Mak et al., 2007; Shingo, 2003). Social and sexual experiences are connected by the hormones involved. The experiments that could be designed comparing variants of each factor on adult neurogenesis are too numerous to list here.

Aging

There is a large body of evidence that demonstrates how aging, without disease, results in a decline in adult neurogenesis. Between 2 and 24 months of age, the number of interneurons in the mouse OB increases significantly, from 1700 to 2400 periglomerular cells (Enwere, 2004). At the same time, the number of BrdU-positive cells is reduced to ~25% of the levels observed at 2 months of age. The decrease in proliferation is observed in both the SVZ and RMS of aging mice (Mobley et al., 2013). Compared to two-month-old animals, aged mice have deficits in the ability to discriminate between two similar odors, related to the loss of new neurons (Enwere, 2004). Thus olfactory discrimination can be used as a read-out to assess adult neurogenesis.

In the SVZ, a number of cytoarchitectural changes occur with aging, though whether they are the cause or consequence of decreased neurogenesis is not known. In middle-aged mice (10–13 months), a thinning of the SVZ occurs that brings the striatum into contact with the ependyma of the lateral ventricle (Luo et al., 2006). The thinning is accompanied by a significant loss of Type C progenitors and Type A neuroblasts. The number of NSCs in middle-aged mice are retained in old age, but there is an increase in the number of NSCs interpolated in the epenydma, which have more than one cilia, as if they are becoming more like ependymal cells (Luo et al., 2006).

Many factors have been investigated and found to have an age-associated effect on adult neurogenesis. A comparison of protein expression shows a decrease in both the epithelial growth factor receptor

(EGFR) and the transforming growth factor-α (TGF-α) in the aging SVZ. A mouse model deficient in TGF-α has a similar reduction in adult neurogenesis, demonstrating that TGF-α is necessary to maintain normal levels of adult neurogenesis. Several studies have experimented with artificially increasing levels of growth factors that restores levels of adult neurogenesis to youthful levels. Increasing the levels of either fibroblast growth factor (FGF)-2 or insulin-like growth factor (IGF)-1 increases proliferation in aging animals (Kuhn et al., 1997; Lichtenwalner et al., 2001). In the aging dentate gyrus, similar decreases in adult neurogenesis occur and are attributed to decreasing proliferation of the NSCs. Decreasing the levels of corticosteroids (adrenal stress hormones known to impair memory with aging) by adrenalectomy, can increase proliferation significantly. Access to a running wheel has also been shown to offset the detrimental effects of social isolation on adult neurogenesis and improves spatial cognition (Huang et al., 2018; Kannangara et al., 2011). Another study found that creating *parabiosis* between young and old mice altered levels of BrdU- and doublecortin-positive cells, with old mice showing increases and young mice showing decreases (Fig. 8.2; Villeda et al., 2011). These experiments demonstrate that NSCs in an aging animal have the potential to proliferate at levels observed in young animals, but that anatomical changes and a decline in factors that regulate neurogenesis are altered with age.

The loss of olfactory ability with aging is not uncommon, nor is it a trivial matter. Deficits in olfaction are as common as one in four persons older than 60 years and approximately half of all people older than 80 years (Doty et al., 1984; Murphy et al., 2002). In rodents, the value of olfaction is indicated by the fact that olfactory bulbectomized rats produce a model of depression (Kucerova et al., 2012; Rinwa et al., 2013; Slotkin and Seidler, 2006; Song and Leonard, 2005). Aside from quality-of-life issues (flavor, aromas), loss of olfaction hinders the detection of dangers such as spoiled food, smoke or natural gas. In the elderly, loss of olfaction is correlated with the risk of malnourishment that accompanies complaints that food is flavorless (Boyce, 2006). Indeed, the elderly require twice as much salt to detect its addition compared to young adults (Stevens et al., 1991).

Recent reports found that loss of olfaction is also predictive of five-year mortality. In a study of more than 3000 people, ages 57–85 years, an odor identification test was given and subjects were classified as normosmic, hyposmic or anosmic. Five years later, cause of death and related records were examined (Pinto et al., 2014). Both anosmia and hyposmia were strong predictors of mortality, with hyposmia being a greater indicator of five-year mortality than nutritional status, cognitive function, mental health, smoking, alcohol use or frailty (Fig. 8.3). Another study compared factors such as education, ethnicity, language, comorbidity, dementia, depression, alcohol use, head injury and smoking to olfactory ability (Devanand et al., 2015). Of the 1169 participants, 369 died during the

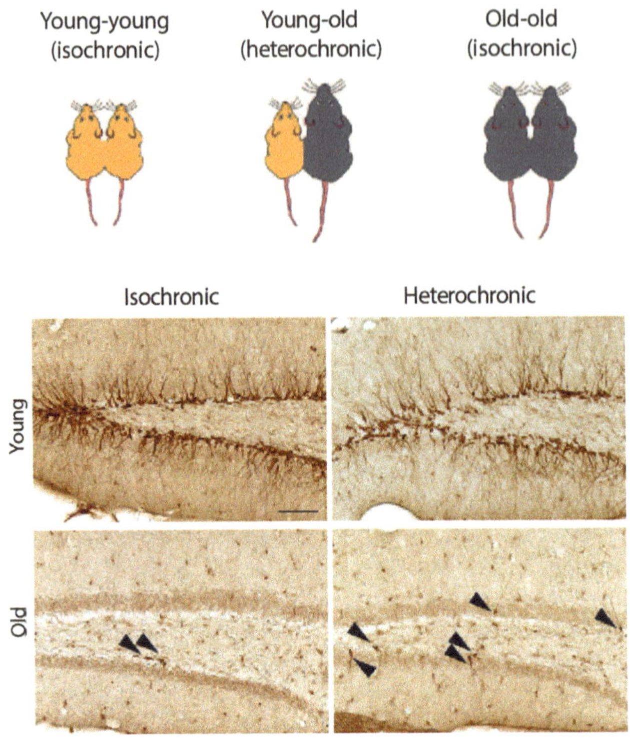

FIG. 8.2 Parabiosis, resulting from the surgical joining of two organisms, alters adult neurogenesis in an age-dependent manner. (Top) An illustration of the parabiotic pairings. (Bottom) In a young brain from a young-young isochronic pairing, doublecortin (DCX) labels a large number of immature neurons in the dentate gyrus. In a young brain from a heterochronic pairing there is a significant reduction in DCX-positive cells. In an old brain from an old-old isochronic pairing there are very few DCX-positive cells. An examination of an old brain from a heterochronic pairing found a significant increase in DCX-positive cells. *Reprinted with permission from Villeda, S.A., Luo, J., Mosher, K.I., Zou, B., Britschgi, M., Bieri, G., Stan, T.M., Fainberg, N., Ding, Z., Eggel, A., et al., 2011. The ageing systemic milieu negatively regulates neurogenesis and cognitive function. Nature 477, 90–94.*

study, and 227 died within 5 years of taking the odor identification test. After controlling for each factor recorded, they found that poor olfaction was strongly associated with mortality (Devanand et al., 2015). Further, loss of olfactory abilities is a clinical indicator for neurodegenerative disorders such as Parkinson disease and Alzheimer disease, with loss of olfactory discrimination being a strong indicator of the associated cognitive decline (Doty, 2012; Sohrabi et al., 2012; Velayudhan, 2015). Therefore investigations on adult neurogenesis and olfactory abilities has important relevance to health, disease and mortality.

Life expectancy increases have intensified the negative impacts of aging on cognitive functions, such as hippocampus-dependent memory (spatial learning and episodic memory). The production of adult-born

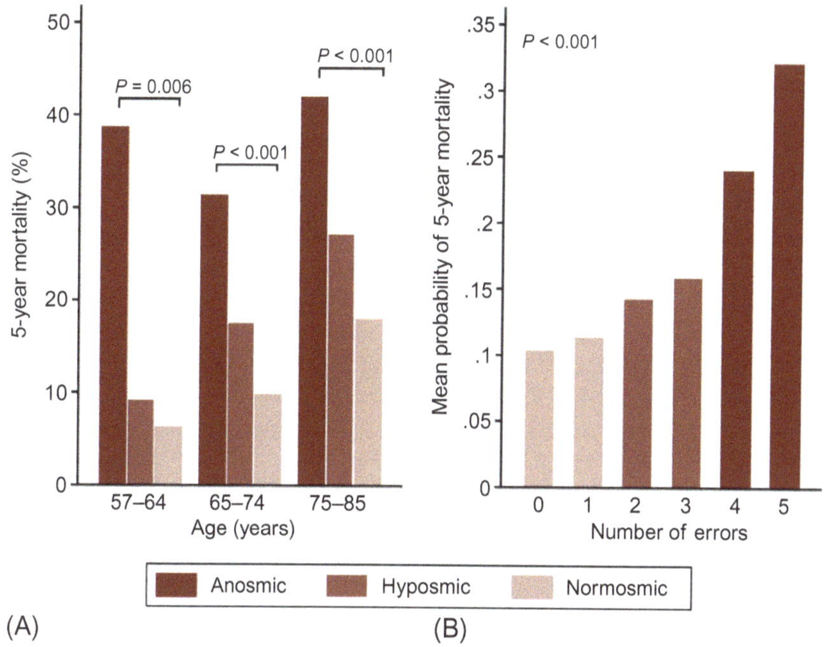

FIG. 8.3 Olfactory dysfunction predicts five-year mortality. (A) The percentage of deceased participants 5 years after taking an odor identification test. (B) A logistic regression with the number of odor identification errors shows a significant increase in the likelihood of five-year mortality. *Image reprinted from Pinto, J.M., Wroblewski, K.E., Kern, D.W., Schumm, L.P., McClintock, M.K., 2014. Olfactory dysfunction predicts 5-year mortality in older adults. PLoS ONE 9, e107541; under the terms of the Creative Commons Attribution License (CC BY). Copyright: © 2014 Pinto et al.*

neurons plays an important role in learning and memory; thus a decrease in adult neurogenesis is an important factor to consider in age-related cognitive decline. With aging there can be as much as a tenfold reduction in the production of adult-born cells in rodent dentate gyrus (Kempermann et al., 2002; Kuhn et al., 1996). Studies on aging rodents have also reported altered baseline activity of hippocampal circuits, abnormal calcium activity, decreased levels of neurotrophic factors, a continuous reduction of Type 1 NSCs and an increase in glucocorticoid signaling (Chapman et al., 2012; Encinas and Sierra, 2012; Patrylo et al., 2007; Talmi et al., 1993; Thibault et al., 2001). An understanding of the mechanisms that regulate proliferation, survival and integration of adult-born neurons may enable us to prevent age-related decreases in cognition. Studies in rodents have already shown that environmental enrichment and running can increase proliferation, suggesting that there may be things individuals can do to prevent aging-dependent decreases in adult neurogenesis (Kannangara et al., 2011; Kempermann et al., 2002).

STATES OF DISEASE

Depression

Major depressive disorder afflicts more than 6% of adults in the United States as of 2016, but can affect as many as 12%–17% of individuals at some point over their lifetime (Kessler et al., 1994; www.nimh.nih.gov). Approximately 40%–50% of the risk for depression is genetic, and is often associated with stress or anxiety (Nestler et al., 2002).

Pharmacological treatments for depression generally target a class of neurotransmitters called monoamines, which includes dopamine (DA), serotonin (5-hdroxytryptamine; 5HT) and norepinephrine (NE). Because DA has not been shown to have a prominent role in depression, treatment primarily targets 5HT and NE neurotransmission with four classes of drugs: tricyclic antidepressants (TCAs), monoamine oxidase (MAO) inhibitors, selective serotonin reuptake inhibitors (SSRIs) and selective serotonin and norepinephrine reuptake inhibitors (SNRIs). The difference between TCAs and SNRIs is that TCAs also act on muscarinic acetylcholine, alpha 1 adrenergic and histamine H1 receptors, and therefore can have unwanted side effects. Another less commonly used drug to treat depression and other mental illnesses is lithium. While the exact mechanism of its effectiveness is not known, it also appears to act upon the monoaminergic signaling system (El Khoury et al., 2001).

TCAs, SSRIs and SNRIs have a common mechanism of action at the synapse. After neurotransmitter is released into the synaptic cleft, the excess neurotransmitter that is not bound by the receptors is taken back into the presynaptic cleft by transport proteins, a process called reuptake (Fig. 8.4). Back in the presynaptic terminal, the neurotransmitter may be recycled into synaptic vesicles or broken down into smaller molecular components. TCAs, SSRIs and SNRIs all act as inhibitors on the transport proteins. The result is that more neurotransmitter remains in the cleft and can continue to bind to receptors on the postsynaptic cell, thereby increasing its excitation. MAO inhibitors act within the presynaptic terminal to prevent a protein called monoamine oxidase from breaking down the neurotransmitters. The result is that the amount of recycled neurotransmitter is increased, and upon depolarization, the presynaptic terminal will release more neurotransmitter.

One caveat to antidepressant drugs is that all of them require prolonged use before any effect is detected. Often several weeks of use is required for the drug to be effective (Duman et al., 2012). This means that simply increasing neurotransmission of 5HT and NE neurons is not the direct cause of relief from depression.

Stress and depression have a detrimental effect on the hippocampus including atrophy of dendritic processes and loss of neurons (Duman et al., 1999; McEwen, 1999). In humans, a loss of hippocampal volume

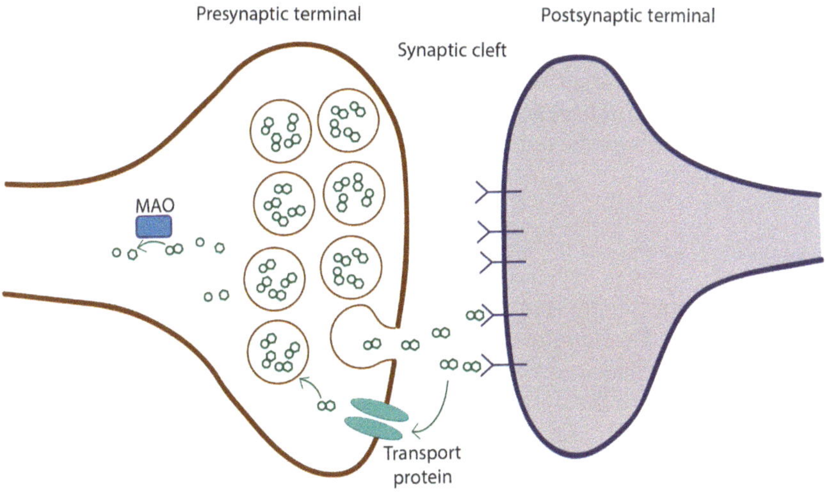

FIG. 8.4 Antidepressant effects on the synapse. When neurotransmitter is released into the synaptic cleft, the remaining molecules that do not bind postsynaptic receptors are taken back into the presynaptic terminal by transport proteins (reuptake). SSRIs, SNRIs and TCAs block re-uptake in order to increase the amount of neurotransmitter available to bind to postsynaptic receptors. In the presynaptic terminal, MAO enzymes break down neurotransmitter into molecular components. MAO inhibitors block this action in order to increase the amount of neurotransmitter available to be packaged into vesicles within the presynaptic terminal, and subsequently released into the synaptic cleft. These drugs all increase the likelihood of postsynaptic receptor activation.

can be observed with brain-imaging techniques in those suffering from long-term depression, posttraumatic stress disorder (PTSD) and other stress-related psychiatric illnesses (Sapolsky, 1996; Sheline et al., 1996). Depression-induced changes to the hippocampus can affect declarative learning and memory (Austin et al., 2001). These changes are relevant to humans because one of the symptoms used to diagnose depression, and many other affective and anxiety disorders, is impaired cognition (Miller and Hen, 2015). From these observed changes, experiments were designed to test the effects of antidepressant drugs on adult neurogenesis.

In rodents, one study compared electroconvulsive shock (ECS), tranylcypromine (TCP, a MAO inhibitor), reboxetine (SNRI) and fluoxetine (SSRI), while another tested lithium (antimanic) on adult neurogenesis. All of the antidepressant treatments increased proliferation in the dentate gyrus, some as rapidly as 2h after some of the treatments, as measured by BrdU (Chen et al., 2002; Malberg et al., 2000). Four weeks later there was a significant increase in cell survival only present in animals treated with the antidepressants (Malberg et al., 2000). Indeed, ablation of adult neurogenesis accompanied by acute stress increases susceptibility to depression (Snyder et al., 2011). However, while ablation of adult neurogenesis does not cause depression, it does eliminate the behavioral responses to fluoxetine and affects cognition (Petrik et al., 2012; Santarelli, 2003). Simply boosting levels of proliferation does not alleviate symptoms of depression, though it may help *prevent* depression (Miller and Hen, 2015; Petrik et al., 2012). Thus while adult neurogenesis appears to be part of the neural circuitry affected by depression it is not solely responsible for mood or affect.

Stress

Stress is a normal, motivating factor in life. However, chronic, long-term stress can have detrimental effects on the brain. Glucocorticoids are adrenal steroids secreted in response to stress, and are essential for maintaining granule cells in the dentate gyrus (Gould et al., 1990). Adrenalectomy in the adult rat eliminates glucocorticoids and results in massive cell death of granule cells (Gould et al., 1990). Antidepressant treatment is reported to increase adult neurogenesis in the dentate gyrus most consistently in rodent models of stress when glucocorticoid signaling is active (David et al., 2009). In a mouse model of stress (chronic restraint), ECS increased both neurogenesis and survival of adult-born cells in the dentate gyrus (Olesen et al., 2017). However, when excess amounts of glucocorticoids, such as those occurring with chronic stress were present, the complexity and length of CA3 pyramidal cell dendrites were decreased, which is suggestive of degeneration (Woolley et al., 1990).

To counteract stress-induced decreases in adult neurogenesis, one study compared whether running, fluoxetine or duloxetine (SNRI) had any beneficial effect (Marlatt et al., 2010). The study used an open field test, where time spent between the open field or near the boundaries (a sign of anxiety) is measured. They found that running was more effective for decreasing anxiety, and it increased proliferation in the dentate gyrus (BrdU-positive cells). The authors concluded that the anxiolytic effects of fluoxetine and duloxetine are likely not mediated by adult neurogenesis. Similar to depression, anxiety has a negative effect on cells of the hippocampus, but it has not been demonstrated that simply increasing adult neurogenesis can lessen anxiety.

The dentate gyrus receives serotonergic innervation from neurons in the raphe nucleus (Fig. 8.5; Vitalis et al., 2013). Inhibition or ablation of these neurons decreases adult neurogenesis and spine density of granule cells; however, one study showed no effect of 5HT administration alone on adult neurogenesis (Huang and Herbert, 2005; Jha et al., 2006; Yan et al., 1997). Recently, a new protein from the c-Jun N-terminal kinase (JNK) family has emerged that may prove to be a potential target for therapeutic regulation of anxiety. JNKs are known to regulate apoptosis through two mechanisms: the activation of transcription factors that regulate apoptotic genes, and direct activation by the phosphorylating

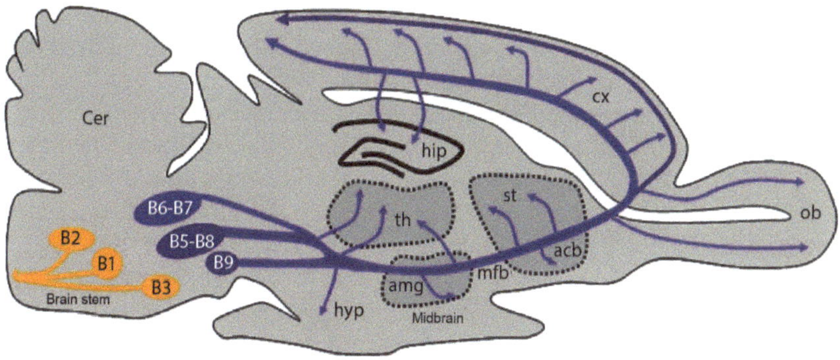

FIG. 8.5 Innervation of the rodent neurogenic niches by 5HT. The 5HT neurons are distributed in nine groups (B1–B9). The posterior groups B1–B4 (yellow, B4 not shown) are located in the midpons to caudal medulla in the raphe magnus; these project to the medulla, spinal cord and the periphery. The remaining 5HT nuclei are in the midbrain and pons and innervate regions in the midbrain and forebrain. B6 and B7 make up the dorsal raphe nucleus and specifically innervate the hippocampus (hip) and OB. B5 and B8 make up the median raphe nucleus. B9 is called the supralemniscal cell group; these cell groups primarily project to the cortex. Light and dark blue arrows, corresponding to the 5HT nucleus color, indicate regions highly innervated by 5HT afferents: the hypothalamus (hyp), amygdala (amg), thalamus (th), accumbens (acc), striatum (st), and cerebral cortex (cx). Cer, cerebellum. *Image reprinted from Vitalis, T., Ansorge, M.S., Dayer, A.G., 2013. Serotonin homeostasis and serotonin receptors as actors of cortical construction: special attention to the 5-HT3A and 5-HT6 receptor subtypes. Front. Cell. Neurosci. 7; under the terms of the Creative Commons Attribution License (CC BY). Copyright © 2013 Vitalis, Ansorge and Dayer.*

pro- and anti-apoptotic proteins that results in apoptosis (Dhanasekaran and Reddy, 2008). Knockout mice missing the *jnk1* gene (JNK1−/−) exhibit increased proliferation and survival of adult-born cells in the dentate gyrus (Mohammad et al., 2018). The JNK1−/− mice displayed lower anxiety in an open field test along with other measures that indicate a "grittier" mouse. In wildtype mice, chronic inhibition of JNK1 by hippocampal infusion or delivery of a retrovirus, showed the specific effect of JNK1, increasing adult neurogenesis and dendritic arborization of immature neurons, while decreasing anxiety in behavioral tests (Mohammad et al., 2018). These data have important implications for psychiatric disorders, as it was recently reported that sequence variation in the *mkk7* gene, an upstream regulator of the JNK signaling pathway, is associated with schizophrenia (Winchester et al., 2012). Thus efforts to understand the link between adult neurogenesis, anxiety, depression and other affective disorders continues to reveal connections.

Stroke

For several weeks after a stroke, neural stem cells in the SVZ give rise to a large number of neuroblasts that migrate to the site of *ischemic* damage

(Arvidsson et al., 2002; Jin et al., 2001; Parent et al., 2002; Zhang et al., 2004). Many experimental stroke models inflict injury using various methods on different brain regions, and have demonstrated induced cell proliferation and expansion of the dentate gyrus and SVZ, and have also noted cells migrating from the SVZ to the sight of injury (Fig. 8.6; Gotts and Chesselet, 2005; Jin et al., 2001, 2015; Parent et al., 2002; Tonchev and Yamashima, 2006). The dentate gyrus is sensitive to ischemic insults that result in learning and memory deficits, but it also shows increased proliferation after a stroke (Lu et al., 2017). However, only a small number of cells undergo neuronal differentiation at the damaged area, prompting studies to determine how a cell's neuronal fate can be altered and integration enhanced, in an effort to improve stroke recovery (Arvidsson et al., 2002).

One of the prevalent models for studying stroke is the mid-cerebral artery occlusion, in which the artery (on one side) is obstructed for specific

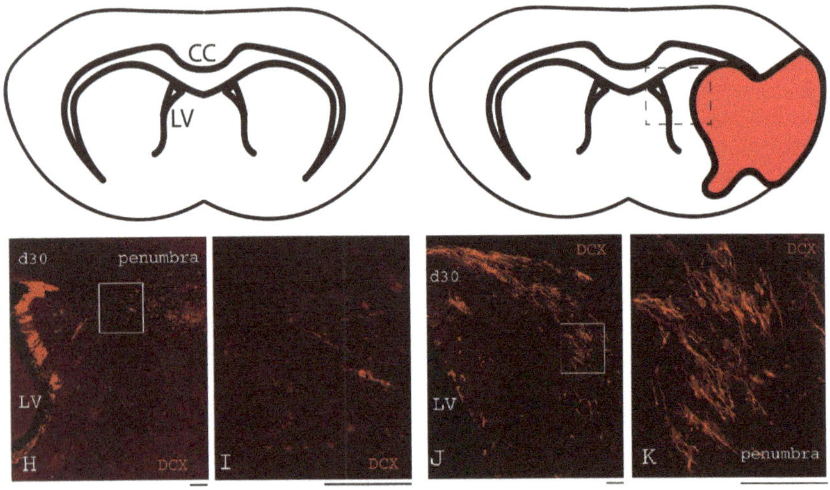

FIG. 8.6 Stroke-induced increases in proliferation and neuroblast migration. (Top left) A coronal section through the rodent brain showing the lateral ventricles (LV) below the corpus callosum (CC). (Top right) A mid-cerebral artery occlusion (MCAO) leads to ischemic cell death (red region) in the striatum as well as surrounding regions in the frontal, parietal and temporal lobes; additional involvement of other regions such as the occipital cortex is dependent on how long the artery is occluded. The MCAO can produce variable damage in the thalamus, substantia nigra and hypothalamus. Dashed box is the region shown in the bottom panel. (Bottom) Proliferation in the SVZ after MCAO on the contralateral (two left images) and ipsilateral (two right images) side of the brain. Boxed regions in H and J are magnified in I and K, respectively. On the ipsilateral side a large number of DCX-positive cells are present that are migrating toward the damaged area. Scale bars = 100 μm. *Bottom images reprinted from Jin, Y., Raviv, N., Barnett, A., Bambakidis, N.C., Filichia, E., Luo, Y., 2015. The Shh signaling pathway is upregulated in multiple cell types in cortical ischemia and influences the outcome of stroke in an animal model. PLoS ONE 10, e0124657 with permission under the terms of the Creative Commons Attribution License (CC BY). Copyright: © 2015 Jin et al.*

time periods, and then released to assess recovery processes (Fig. 8.6). This method results in damage to the striatum, a structure close to the SVZ, as well as surrounding brain regions in one hemisphere (Carmichael, 2005). SVZ-born cells migrate to the striatum where survival and integration can be studied. The number of stroke-induced, adult-born neurons is reduced by about 80% 2 weeks after being born (Arvidsson et al. 2002). Adult-born cells in the *penumbra* must accomplish fate specification, migration, survival, maturation and integration in an inflammatory environment where much change is occurring. Several factors have been examined in rodent stroke models to assess their effect on migration, survival and/or integration. Manipulating factors such as Shh and Wnt signaling, anti-inflammatories, chemokines, cannabinoids and cytokines have all resulted in various levels of improvement (Cunningham et al., 2012; Ducruet et al., 2012; Ekdahl et al., 2009; Galve-Roperh et al., 2007; Iosif et al., 2008; Jin et al., 2015; Shruster et al., 2012; Wang et al., 2012b).

Neuroblast migration to the injured regions is regulated by the chemokine CXCL12, a protein important for HSC homing in the HSC niche (Khurana et al., 2014). Activated microglia and reactive (inflamed) astrocytes at the site of injury express CXCL12 that attracts the CXCR4-expressing neural progenitors (Robin et al., 2006). The endothelial cells also express CXCL12 to promote chain migration toward the damaged area (Ohab et al., 2006).

The inflammatory nature of the ischemic area does not promote neuronal survival, and anti-inflammatory treatments can increase the number of striatal neuroblasts and mature neurons in the dentate gyrus while decreasing the number of activated microglia (Hoehn et al., 2005; Liu et al., 2007). The relationship between new neuroblasts and microglia is complex, with studies showing beneficial and detrimental effects of the inflammatory response (for review see Lindvall and Kokaia, 2015).

One of the goals of adult neurogenesis research is to determine how adult-born cells can replace the neurons lost to a stroke. In rodents, two studies reported that striatal projection neurons were produced by adult-born cells following striatal ischemia (Arvidsson et al., 2002; Parent et al., 2002). However, other studies have reported that only interneurons are produced from adult-born cells, which they hypothesized may be due to the regional heterogeneity of the SVZ (Collin et al., 2005; Liu et al., 2009; Merkle et al., 2007; Teramoto et al., 2003). Another study in rats showed the successful integration of cholinergic and GABAergic adult-born cells after a stroke (Hou et al., 2008). While these data support endogenous repair by adult-born neurons, the mechanisms of stroke-induced increases in proliferation that determine fate specification, differentiation and integration remain unknown. Furthermore, features of the environment that will support neuroblast survival are also not well understood.

One alternative solution to regulating endogenous repair is to transplant stem cells into the injured area. Although there is not yet evidence that stem cells replace lost neurons, several clinical trials have been carried out, providing important information on the safety of placing stem cells in the human brain (Wang et al., 2017). As of yet, there has been no evidence of tumor formation and several studies resulted in improved patient outcomes due to neuroprotective factor secretion by stem cells, or stem cells acting as scaffolding for repair mechanisms (Banerjee et al., 2014; Bhasin et al., 2013; Qiao et al., 2014). Clearly, much more work is required before clinically relevant solutions are available on a large scale.

Epilepsy

Epilepsy is a diagnosis given to a group of neurological disorders that share seizure activity as a central symptom. Temporal lobe epilepsy (TLE) is the most common form of epilepsy in adults, in which seizures are initiated in the temporal lobe but can spread throughout the brain. One of the symptoms of TLE is cognitive dysfunction (Elger et al., 2004). Often epilepsy is precipitated by a head injury that can occur several months to years prior to the symptomatic onset of the disorder (Cho et al., 2015). Some of the changes that can occur between injury and onset are reactive gliosis, inflammation and neurogenesis (Blümcke et al., 1999).

The effect of seizure activity on the dentate gyrus is aberrant neurogenesis. After a seizure there is increased proliferation that produces more type 2 cells, and the integration of ectopically placed granule cells in the hilus that have an additional basal (hilar) dendrite (Fig. 8.7; Jessberger and Parent, 2015; Parent et al., 1997). Mossy fiber sprouting and neuronal hypertrophy are also observed. The basal dendrite receives excitatory input from mossy fibers that could lead to increased excitation (Jessberger and Parent, 2015).

It is unclear if aberrant neurogenesis contributes to the etiology of epilepsy. One study used a genetic method to eliminate dividing neural stem/progenitor cells prior to initiating seizure activity in mice (Cho et al., 2015). Four weeks after the ablation they injected pilocarpine in order to generate status epilepticus (SE), a state of continuous seizure activity that serves to develop epilepsy in the animal. Pilocarpine is a drug that acts on muscarinic receptors to shift the balance of excitation and inhibition in the brain. SE is halted by a benzodiazepine called diazepam, also known as Valium (Curia et al., 2008). The outcome of the experiment was an elimination of aberrant neurogenesis that decreased the number of spontaneous seizures, which rescued the mice from the typical cognitive decline. While the severity of spontaneous recurring seizures did not change, the reduction in frequency lasted up to a year. The authors hypothesized that preventing the production of ectopic granule cells, which receive an abun-

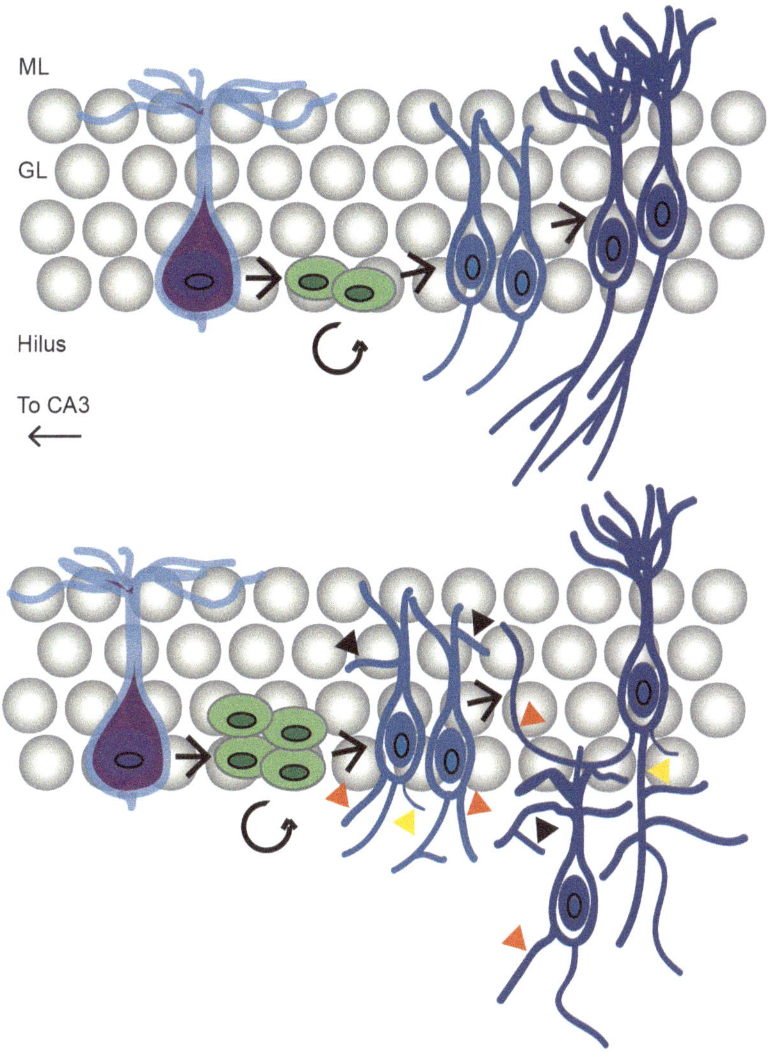

FIG. 8.7 Aberrant neurogenesis in the epileptic dentate gyrus. (Top) Adult neurogenesis gives rise to immature (light blue) and mature (dark blue) granule cells that extend an axon through the hilus to CA3. (Bottom) In the epileptic dentate gyrus, there is increased proliferation, and the new neurons show unusual dendritic arborization (black arrowheads), an additional basal dendrite (red arrowheads) and axonal process (yellow arrowheads). New adult-born granule cells mature ectopically in the hilus and introduce connectivity with mossy fibers (not shown) that leads to increased excitation.

dance of excitatory inputs from mossy fibers in the hilus, was responsible for the improved outcome (Cho et al., 2015). Thus targeting aberrant neurogenesis could be a rich area of research leading to future epilepsy treatments.

Degeneration

Alzheimer Disease

Alzheimer disease (AD) is the most common neurodegenerative disorder, estimated to affect close to six million people age 65 years and older. AD is characterized by the loss of neurons, most prevalently in the hippocampus, entorhinal cortex and neocortex. The classic theory is that the degeneration is due to the accumulation of neurofibrillary tangles made up of abnormally hyperphosphorylated cytoskeletal tau protein and β-amyloid (Aβ) plaques, a fragment of amyloid precursor protein (APP), which is a type 1 *integral membrane protein*. Several proteins implicated in AD pathology influence adult neurogenesis.

APP is a protein that spans the cell membrane and is cleaved at several locations by one of three multisubunit protease complexes: α-secretase, β-secretase and γ-secretase (Fig. 8.8). The APP cleavage sights and the fragments produced can be categorized into the amyloidogenic and nonamyloidogenic pathway. Typically, α-secretase cleaves APP producing the non*amyloidogenic,* soluble APPα (sAPPα), a membrane-bound carboxyl-terminal fragment, and the C-terminal fragment (CTF) 83. The intramembrane γ-secretase complex cleaves APP at the C-terminus and produces p3 and the APP intracellular domain (AICD) fragments. A role for p3 is not known; however, AICD enters the nucleus where, with other proteins, it transcriptionally activates the expression of genes such as GSK3β and EGFR (Chow et al., 2010). When β-secretase acts together with γ-secretase, the result is the sAPPβ and Aβ fragments, constituting the amyloidogenic pathway. Presenilins (PS) are the catalytic components of γ-secretase, and mutations in PS1 and PS2 favor the amyloidogenic pathway, increasing production of Aβ.

The biggest risk factor for AD is advanced age, with the majority of cases occurring sporadically. However, the ε4 allele in apolipoprotein E (ApoE4), a glycoprotein that transports cholesterol in the blood and plays a role in cellular repair, is a known genetic risk factor as well as a few rare mutations in the genes encoding APP and presenilins that increase the likelihood of AD with age.

The hippocampus and OB are two of the first locations where plaques and tangles are observed (Braak et al., 1993; Ohm and Braak, 1987). Many of the molecules in the amyloidogenic pathway responsible for AD pathology have roles in hippocampal adult neurogenesis. For example, inhibition of PS1 gene expression enhances γ-secretase-dependent neuron differentiation (Gadadhar et al., 2011). The effect is likely due to decreased AICD production, which can decrease progenitor cell differentiation in the dentate gyrus through a pro-inflammatory mechanism (Ghosal et al., 2010). Studies on three different PS1 mutations reported a negative effect on adult neurogenesis, either through decreased proliferation,

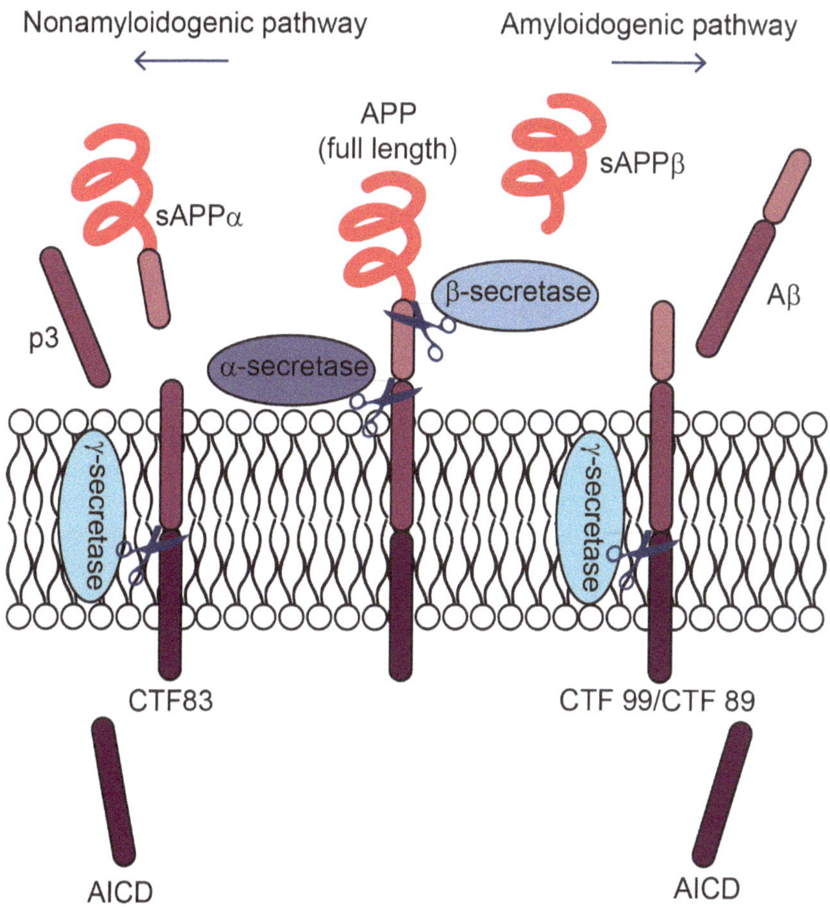

FIG. 8.8 APP processing by secretase activity. In the nonamyloidogenic pathway α-secretase first cleaves full length APP into sAPPα and CTF83. The γ-secretase complex further cleaves the membrane-bound CTF83 into AICD and p3. In the amyloidogenic pathway β-secretase first cleaves full length APP into sAPPβ and CTF99/CTF 89. The γ-secretase complex further cleaves CTF 99/CTF 89 into Aβ (potentially toxic) and AICD.

differentiation or survival (Choi et al., 2008; Rodríguez et al., 2008; Verret et al., 2007). Presenilins also cleave other type 1 integral membrane proteins such as the Notch receptor, which makes discerning its role in adult neurogenesis more challenging (De Strooper et al., 2012).

Complicating this story are two reports that found increased adult neurogenesis in human brains with probable AD that was confirmed by neuropathology (Jin et al., 2004a). The postmortem brains had increased expression of neurogenesis markers in the dentate gyrus compared to controls (Jin et al., 2004a, b). The finding is consistent with increased adult neurogenesis after ischemia in rats, but not with AD studies in mice. Next, the authors used mice to express three APP mutations inserted under the platelet-derived growth factor-β (PDGF) promoter (Jin et al., 2004b). This mouse model also had increased adult neurogenesis in both the dentate gyrus and SVZ. Interestingly, at 1 year, when proliferation begins to decline with normal aging, the triple transgenic mouse had increased proliferation (Jin et al., 2004b; Mobley et al., 2013). While the human tissue samples were from confirmed AD cases, it is not known whether the subjects suffered from a sporadic form of AD or if they carried the same mutations used in the animal study (Jin et al., 2004a, b). Conversely, in a conditional double knockout of PS1 and PS2, neurodegeneration in the late stages of AD was offset by increased adult neurogenesis at early stages, suggesting that targeting adult neurogenesis could be a way to delay the onset of AD symptoms (Chen et al., 2008).

Other mouse models of AD show detrimental effects from molecules implicated in AD. Aβ fragments are toxic to neural progenitors, with increased apoptosis due to calcium dysregulation in both the hippocampus and the SVZ (Haughey et al., 2002a, b). Mice with increased expression of Aβ40 and Aβ42-43 (molecules produced due to different cleavage sites of APP by β-secretase) in cortical and limbic regions of the brain, including the hippocampus, had deficits in learning and memory (Hsiao et al., 1996). ApoE4 elimination can increase NSC proliferation while depleting the Type 1 pool, suggesting a dual role in stem cell maintenance (Yang et al., 2011). Thus the molecules thought to be responsible for AD neurodegeneration contribute in various ways to the regulation of adult neurogenesis.

One of the earliest indicators of cognitive impairment is the loss of one's sense of smell (Doty, 2012; Sohrabi et al., 2012; Velayudhan, 2015). In a longitudinal study, called the Rush Memory and Aging Project, an odor identification test was given to participants, followed by assessment of neuropathy at the time of death (Wilson et al., 2007). At the time of publication more than 1000 participants had enrolled and of the 166 that were deceased (mean age of death = 87.5), 129 had been assessed for neuropathy. Correlating performance on the odor identification test and neuropathy, the authors reported that the presence of tangles or the burden of amyloid was inversely correlated with odor identification performance. The most significant effect was the presence of tangles in the hippocampus, entorhinal cortex and inferior temporal cortex (Wilson et al., 2007). Another study compared elderly, nondemented adults to those with AD and confirmed that those with AD performed significantly lower on an odor identification test (Suzuki et al., 2004). Within the AD group, those carrying one or two ApoE4 alleles had the strongest correlation. Another study analyzed the effect of a mutation in APP and PS1 and found that the number of somatostatin- and calretinin-positive cells in the OB were decreased in transgenic animals, suggesting a particular vulnerability that was correlated with the presence of Aβ plaques, while other interneuron populations seemed unaffected (De la Rosa-Prieto et al., 2016; Saiz-Sanchez et al., 2013). The fact that PS1 is an important regulator of Notch and Wnt signaling warrants further investigation in to the somatostatin-producing stem cells in the SVZ and how they are affected by AD pathology (Merkle et al., 2007; Winner et al., 2011).

The evidence for stem cell replacement in animal models of AD prompted several clinical trials using UCB-MSCs, which were generally administered intravenously. So far, no adverse effects have been reported, though an inhibition of cognitive decline has not been observed (Kwak et al., 2018). With clinical trials still underway, the promise of stem cell research remains a hope for treating AD.

Parkinson Disease

Parkinson disease (PD) is the second most common degenerative disorder (after AD) and is characterized by the loss of dopaminergic neurons in the substantia nigra, a midbrain nucleus that is part of the basal ganglia, that projects to the striatum. It was first described in 1817 by James Parkinson as a shaking palsy (Goetz, 2011). The symptoms associated with dopaminergic neuron loss in the substantia nigra, are *hypokinesia*, rigidity, tremor and postural instability. It is hypothesized that neuronal degeneration is due to the build-up of abnormal circular structures called Lewy bodies, with a dense core of α-synuclein (αSYN) protein in the cell cytoplasm, as well as α-SYN build-up at the synapses. Lewy body pathology can spread to other areas of the brain including the midbrain, cerebral cortex and areas of the autonomic nervous system (Braak and Del Tredici, 2009; Edwards et al., 1992). There are a few rare point mutations of *α-synuclein*, a gene found in the PARK-1 *locus*, associated with increased risk for the development of PD, and overexpression of αSYN is also a form of familial PD. Nine additional loci have been confirmed as a genetic cause of PD, but most cases of PD appear to be sporadic (Gasser et al., 2011),

The symptoms of PD are thought to occur due to either toxic gain-of-function or toxic loss-of-function conditions that may be due to both genetic and environmental factors. The toxic gain-of-function theory involves a dominant genetic mutation, in this case α-synuclein, which becomes misfolded, and aggregates in the cell, forming Lewy bodies. Toxic loss-of-function is when a recessive gene fails to produce a necessary protein, in this case a gene called *parkin*, which is responsible for tagging defective or misfolded proteins with ubiquitin, a protein that targets the defective proteins to proteasomes where they are broken down. Without Parkin directing the defective proteins for breakdown, they accumulate in the cell, eventually killing it. Dopaminergic neurons of the substantia nigra are especially sensitive to this accumulation, which may be due to their long axons and complex dendrites that require extensive axonal transport for the maintenance of up to ~150,000 synapses

(Braak et al., 2006; Roberts et al., 2002). While it is logical to hypothesize that *parkin* mutations are responsible for αSYN accumulation, there is only one glycosylated form of αSYN that is degraded in relation to Parkin, and experimentally was not a significant interaction (Sulzer, 2007). However, because of Parkin's role in protein degradation, its malfunction is believed to be one of the leading familial causes of PD (Gasser et al., 2011). The prevalence of sporadic PD has led to the idea that multiple factors such as genetic vulnerability, environmental factors such as neuroinflammation caused by pathogens (bacterial or viral infections), dysregulation of inflammatory pathways, *oxidative stress*, ubiquitin-protease dysfunction and aging are involved in the protective sequestration of protein in Lewy bodies, or protein aggregates (Deleidi and Gasser, 2013; Sulzer, 2007).

Changes in dopaminergic innervation and dopamine concentration in the striatum can be detected 20–30 years before motor symptoms manifest. Nonmotor symptoms such as olfactory dysfunction, cardiac sympathetic denervation and constipation, REM sleep disorder, depression and anxiety have been reported as early as years or decades before motor symptoms appear. The anatomy of disease progression finds that the OB and brainstem nuclei are the earliest brain areas affected followed by midbrain, mesocortical and finally cortical regions (Braak et al., 2003). Indeed, hyposmia is one of the earliest premotor symptoms of PD (Doty, 2012). Cognitive impairment has also been reported in the early motor stage of PD (Aarsland et al., 2010). Thus it is not surprising that mouse models of PD show detrimental effects on adult neurogenesis.

In mice, overexpression of *α-synuclein* under the control of the platelet-derived growth factor β (PDGFβ) results in both decreased survival and increased cell death of adult-born neurons in the OB, without affecting proliferation in the SVZ (Masliah et al., 2000). Overexpression of the human A53T α-synuclein mutant causes a decrease in proliferation that is amplified with age, and includes reduced survival of adult-born neurons in the SVZ and dentate gyrus (Kohl et al., 2012; Winner et al., 2011). Studies on postmortem human brains to evaluate the state of adult neurogenesis report conflicting findings: either decreased proliferation or no change in proliferation (van den Berge et al., 2011; Höglinger et al., 2004). The effects of αSYN accumulation or loss of dopaminergic innervation in the neurogenic centers of the brain may contribute to the early nonmotor symptoms of hyposmia and depression.

The role of inflammation is being incorporated into many disease and brain injury processes. Chronic neuroinflammation is present in AD and PD brains of both humans and animals (Cicchetti et al., 2002; McGeer et al., 1988). Not only does αSYN activate (inflame) astrocytes, αSYN can propagate from neurons to astrocytes (Lee et al., 2010; McGeer and McGeer, 2008). Another study found that αSYN induces toxic levels of stress signaling molecules and that αSYN mutations found in humans exacerbated the toxicity (Klegeris et al., 2006, 2008). In a mouse model of PD, anti-inflammatory treatment improved NSC proliferation in the SVZ (Worlitzer et al., 2012). These studies emphasize the need to understand the connection between neurodegeneration, inflammation and adult neurogenesis.

The two most common animal models of PD are produced by the administration of drugs that target dopaminergic neurons, 6-hydroxydopamine (6-OHDA) or 1-methyl-4-phenyl-1,2,3,6-tetrahydropyridine (MPTP). Both result in the degeneration of DA neurons and cause motor phenotypes. Consistent with human studies, mouse MPTP models of PD have shown conflicting results—either detrimental (apoptosis of migrating neuroblasts) or no change in SVZ proliferation or migration (van den Berge et al., 2011; He et al., 2006). Studies using 6-OHDA also have mixed results, showing either decreased or increased proliferation in the SVZ; however, neuronal differentiation also appears compromised in some of the studies (Aponso et al., 2008; Baker et al., 2004; Höglinger et al., 2004; Winner et al., 2006). Another study used the 6-OHDA mouse model to test the effect of the common treatment for PD patients, L-DOPA, the precursor of DA on adult neurogenesis (Chiu et al., 2015). The administration of L-DOPA is thought to increase the concentration of DA at the remaining dopaminergic neurons. In mice, L-DOPA normalized adult neurogenesis in the SVZ and dentate gyrus after an initial 6-OHDA-induced decrease (Chiu et al., 2015). Long-term treatment with L-DOPA also improved measurements of depression and anxiety in the mice (Chiu et al., 2015). Using a regression analysis of the postmortem analyses of human PD brains also suggests that L-DOPA increases adult-neurogenesis in the SVZ (O'Sullivan et al., 2011).

Because PD causes the loss of a specific neuron type, stem cells to replace dopaminergic neurons seems like it would be a straightforward solution, and since the 1980s 300–400 persons have received fetal cell transplants to treat PD (Wang et al., 2017). However, while *open-label* studies were promising, controlled experiment results were highly variable and the incidence of dyskinesia deterred further clinical trials using these methods (Karussis et al., 2013). Furthermore, a postmortem study on the brains of persons with PD that had received fetal grafts, found the presence of Lewy bodies 10–15 years later in the grafted neurons (Kordower et al., 2008; Li et al., 2008).

NSCs were once thought to be a potential source of new neurons for replacement therapy. However, while NSCs isolated from a human PD brain were able to form neurospheres in culture, they were unable to differentiate into neurons without the presence of ESCs, suggesting they were in a compromised state (Wang et al., 2012a). Clinical trials using human NSCs found that only 15% of the cells produced were DA neurons, again deterring further interest in this method (Karussis). Alternatively, mouse fibroblasts can be reprogrammed to become DA neurons, and do alleviate

PD symptoms when transplanted into a mouse model of PD (Kim et al., 2011). In 2017 a study reported that human cells reprogrammed to become DA neurons were implanted in nonhuman primates treated with MPTP (Kikuchi et al., 2017). The neurons survived for 2 years without tumor formation, and they integrated into existing circuitry. Furthermore, the study reported that the primates increased their spontaneous movements compared to nontreated controls. The authors of the study are hoping to begin a clinical trial to test the treatment in humans (Callaway, 2017).

Amyotrophic Lateral Sclerosis

Amyotrophic lateral sclerosis (ALS) is a fatal neurodegenerative disease (third most common after AD and PD) for which there is also currently no cure. Motor neuron degeneration typically begins in the spinal cord, eventually affecting central motor neurons. The cause of death is usually due to respiratory failure, and occurs 2–3 years after diagnosis.

Ninety percent of ALS cases are sporadic, while the remaining 10% of cases are familial. The first gene linked to familial ALS is the Cu,Zn-superoxide dismutase (SOD1), which accounts for 20%–25% of familial cases (Rotunno and Bosco, 2013). Another genetic cause is an expansion repeat of the hexanucleotide GGGGCC in the c9Orf72 gene, in which patients carrying the expansion may have up to 4400 copies of the repeat. Furthermore, the repeat expansion has a higher association with cases of ALS that are accompanied by frontotemporal dementia (FTD), accounting for 23%–50% of familial ALS cases (Ticozzi et al., 2014). Sporadic cases of ALS are thought to arise from a combination of oxidative stress, inflammation and loss of glial neurotrophic support, though misfolded SOD1 may also contribute to sporadic cases (Faravelli et al., 2014; Rotunno and Bosco, 2013). Indeed, metal depletion and oxidation, or posttranslational modifications can induce SOD1 misfolding (Rotunno and Bosco, 2013).

Although the presence of stem cells has been confirmed in the mammalian spinal cord, their regenerative ability in mammals is much less than what is observed in lower vertebrates such as the salamander, other amphibians and fish (Meletis et al., 2008; Tazaki et al., 2017). The central canal of the spinal cord is filled with cerebral spinal fluid and lined with ependymal cells, of which a subset are stem cells with proliferative potential (Hamilton et al., 2009; Li et al., 2016). After a spinal cord injury, the stem cells can divide, and have been observed to respond to injury with up to a 50-fold increase in proliferation as compared to proliferation in an intact spinal cord (Johansson et al., 1999). Ependymal cell-derived progeny migrate toward the site of injury, where they differentiate into astrocytes that contribute to the glial scar formation (Barnabé-Heider et al., 2010; Johansson et al., 1999; Meletis et al., 2008). Thus an understanding of the differences and similarities between spinal cord stem cells of mammalian and nonmammalian species may enable us to reprogram them to produce neurons.

In the SOD1 mouse model of ALS, there is an increase in stem cell proliferation in the cervical, thoracic and lumbar regions at the onset of the disease (70 days old). The stem cells migrate predominantly to the dorsal horn of the spinal cord in the cervical and thoracic regions, but in the lumbar region, stem cells were observed in both the lumbar and ventral regions. (Chi et al., 2007). As the disease progresses, the number of stem cells remains high with a dramatic increase in the lumbar region (115 days old). These data show the proliferative potential of stem cells in the spinal cord, however, reprogramming of the cells or alterations to the environmental/niche factors are required in order for neurogenesis to occur.

An examination of stem cells in the neurogenic centers of the brain in a transgenic SOD1 mouse model of ALS found some differences compared to control mice (Liu and Martin, 2006). In mice, the stages of disease progression are broken down into presymptomatic, symptomatic and late stage (Gurney et al., 1994). In the presymptomatic and symptomatic mice there is a significant reduction in proliferation in the SVZ, RMS, OB, and dentate gyrus (Liu and Martin, 2006). In the SVZ, there is a reduction in stem cell markers that suggests a loss of this population, though markers of Type C progenitors and Type A migrating neuroblasts remain stable throughout symptomatic stages. However, there is a reduction in the number of adult-born neurons in the OB. When NSCs were isolated from the SVZ of SOD1 mice, the capacity to form neurospheres in culture was retained, though reduced compared to NSCs from the SVZ in control mice (Liu and Martin, 2006). NSCs isolated from the dentate gyrus of SOD1, presymptomatic mice expressed more markers of NSCs than control mice. They also produced neurospheres and survived more passages in culture. In contrast, NSCs from symptomatic mice did not produce neurospheres (Lee et al., 2011). Changes in the neurogenic niche are observed as well, with loss of parvalbumin-positive interneurons in the dentate gyrus that are accompanied by increased measures of anxiety and impaired long-term memory (Quarta et al., 2015). In a small study on postmortem brains from humans with ALS ($n=9$ with ALS, $n=4$ controls), the researchers observed an increase in markers of proliferation in the ALS brains, that was even greater in the two brains from patients with ALS-FTD (Galán et al., 2017). In the dentate gyrus they saw a decrease in markers of proliferation. Thus ALS has an effect on NSCs in the SVZ and dentate gyrus, and increasing our understanding of the changes that occur may enable us to protect them.

Stem cell replacement therapy in the spinal cord and motor cortex of ALS patients could be beneficial in two different ways. The first is through the replacement of lost cells, and the second through the release of trophic factors (Faravelli et al., 2014). Studies to assess the effect of stem cells on ALS in animal models is increasing our knowledge on stem cell delivery, survival and integration where they are needed (Teng et al., 2012).

While there is no cure for the degenerative diseases reviewed here, a large body of work in both humans and animals shows the potential of physical, aerobic exercise to protect against the loss of gray matter volume, while improving learning outcomes and delay or even attenuate cognitive decline (for review see Ahlskog et al., 2011). In humans the mechanisms by which exercise affects adult neurogenesis is not clear, though many animals studies show increased expression of brain neurotrophic factors (Kobilo et al., 2011; Marlatt et al., 2012). Thus the benefit of staying physically active cannot be overemphasized!

SUMMARY

Adult neurogenesis in the human SVZ and dentate gyrus has been both confirmed and denied. The early study by Eriksson et al. was recently supported by a study demonstrating that adult neurogenesis produces new striatal neurons in both the SVZ and striatum (Eriksson et al., 1998; Ernst et al., 2014). The 2014 report helped to clear up some confusion about the role of SVZ-produced neurons in the adult human, as a previous study had identified very few migrating neurons in the human RMS and no evidence of new neurons in the human OB (Wang et al., 2011). In the dentate gyrus a debate has been fueled by two papers published in 2018: one stating no evidence of adult neurogenesis in humans older than 18 years, and the other reporting evidence of adult neurogenesis in humans examined between the ages of 14 and 79 years (Boldrini et al., 2018; Sorrells et al., 2018). Interestingly, the study from Boldrini follows a report from many of the same scientists that found that antidepressant treatments increased markers of neural progenitor and dividing cells in human hippocampi (Boldrini et al., 2009). The comparison was on brains from deceased patients diagnosed with depression, of which some had been treated with SSRIs or TCAs (Boldrini et al., 2009). Obviously, additional studies and improved technology are necessary before a final verdict can be given on whether there is adult neurogenesis in humans, and the nature of its functional role.

DEFINITIONS

Amyloidogenic	Producing a starch-like protein that can form aggregates or deposits.
Chemosensory	Relating to the chemosensory systems, taste and smell.
Hypokinesia	Partial or complete loss of muscle movement.
Integral membrane proteins	A protein that is structurally or functionally a permanent or firmly anchored part of a cell's membrane.

Ischemia	A restriction in blood supply resulting in a loss of oxygen and death to the cells.
Locus	A region of DNA consisting of several millions of base pairs.
Open-label	A clinical trial in which both the physicians and the patients know if they are in the treatment or control group.
Oxidative stress	An imbalance between the production of reactive oxygen species (free radicals) and antioxidant defenses. Inflammation can induce oxidative stress.
Parabiosis	"Living beside." The anatomical union of two organisms that develop shared physiological systems. This can result from experimental, surgical means or spontaneously in utero, as with conjoined twins.
Parvalbumin	A calcium-binding protein structurally related to calmodulin, with three EF-hand motifs. It is found in fast-firing nerve and muscle cells.
Penumbra	The anatomical region surrounding the site of an ischemic event.

References

Aarsland, D., Bronnick, K., Williams-Gray, C., Weintraub, D., Marder, K., Kulisevsky, J., Burn, D., Barone, P., Pagonabarraga, J., Allcock, L., et al., 2010. Mild cognitive impairment in Parkinson disease: a multicenter pooled analysis. Neurology 75, 1062–1069.

Ahlskog, J.E., Geda, Y.E., Graff-Radford, N.R., Petersen, R.C., 2011. Physical exercise as a preventive or disease-modifying treatment of dementia and brain aging. Mayo Clin. Proc. 86, 876–884.

Alpass, F.M., Neville, S., 2003. Loneliness, health and depression in older males. Aging Ment. Health 7, 212–216.

Alvarez, D.D., Giacomini, D., Yang, S.M., Trinchero, M.F., Temprana, S.G., Buttner, K.A., Beltramone, N., Schinder, A.F., 2016. A disynaptic feedback network activated by experience promotes the integration of new granule cells. Science 354, 459–465.

Aponso, P.M., Faull, R.L.M., Connor, B., 2008. Increased progenitor cell proliferation and astrogenesis in the partial progressive 6-hydroxydopamine model of Parkinson's disease. Neuroscience 151, 1142–1153.

Arvidsson, A., Collin, T., Kirik, D., Lindvall, O., 2002. Neuronal replacement from endogenous precursors in the adult brain after stroke. Nat. Med. 8, 963–970.

Austin, M.P., Mitchell, P., Goodwin, G.M., 2001. Cognitive deficits in depression: possible implications for functional neuropathology. Br. J. Psychiatry 178, 200–206.

Baker, S.A., Baker, K.A., Hagg, T., 2004. Dopaminergic nigrostriatal projections regulate neural precursor proliferation in the adult mouse subventricular zone. Eur. J. Neurosci. 20, 575–579.

Banerjee, S., Bentley, P., Hamady, M., Marley, S., Davis, J., Shlebak, A., Nicholls, J., Williamson, D.A., Jensen, S.L., Gordon, M., et al., 2014. Intra-arterial immunoselected CD34+ stem cells for acute ischemic stroke: CD34+ stem cells for acute ischemic stroke. Stem Cells Transl. Med. 3, 1322–1330.

Barker, J.M., Galea, L.A.M., 2008. Repeated estradiol administration alters different aspects of neurogenesis and cell death in the hippocampus of female, but not male, rats. Neuroscience 152, 888–902.

Barnabé-Heider, F., Göritz, C., Sabelström, H., Takebayashi, H., Pfrieger, F.W., Meletis, K., Frisén, J., 2010. Origin of new glial cells in intact and injured adult spinal cord. Cell Stem Cell 7, 470–482.

Bhasin, A., Padma Srivastava, M.V., Mohanty, S., Bhatia, R., Kumaran, S.S., Bose, S., 2013. Stem cell therapy: a clinical trial of stroke. Clin. Neurol. Neurosurg. 115, 1003–1008.

Blümcke, I., Beck, H., Lie, A.A., Wiestler, O.D., 1999. Molecular neuropathology of human mesial temporal lobe epilepsy. Epilepsy Res. 36, 205–223.

Boldrini, M., Underwood, M.D., Hen, R., Rosoklija, G.B., Dwork, A.J., John Mann, J., Arango, V., 2009. Antidepressants increase neural progenitor cells in the human hippocampus. Neuropsychopharmacology 34, 2376–2389.

Boldrini, M., Fulmore, C.A., Tartt, A.N., Simeon, L.R., Pavlova, I., Poposka, V., Rosoklija, G.B., Stankov, A., Arango, V., Dwork, A.J., et al., 2018. Human hippocampal neurogenesis persists throughout aging. Cell Stem Cell 22, 589–599.e5.

Boyce, J.M., 2006. Effects of ageing on smell and taste. Postgrad. Med. J. 82, 239–241.

Braak, H., Del Tredici, K., 2009. Neuroanatomy and pathology of sporadic Parkinson's disease. Adv. Anat. Embryol. Cell Biol. 201, 1–119.

Braak, H., Braak, E., Bohl, J., 1993. Staging of Alzheimer-related cortical destruction. Eur. Neurol. 33, 403–408.

Braak, H., Del Tredici, K., Rüb, U., de Vos, R.A.I., Jansen Steur, E.N.H., Braak, E., 2003. Staging of brain pathology related to sporadic Parkinson's disease. Neurobiol. Aging 24, 197–211.

Braak, H., Bohl, J.R., Müller, C.M., Rüb, U., de Vos, R.A.I., Del Tredici, K., 2006. Stanley Fahn Lecture 2005: the staging procedure for the inclusion body pathology associated with sporadic Parkinson's disease reconsidered. Mov. Disord. 21, 2042–2051.

Brown, J., Cooper-Kuhn, C.M., Kempermann, G., Van Praag, H., Winkler, J., Gage, F.H., Kuhn, H.G., 2003. Enriched environment and physical activity stimulate hippocampal but not olfactory bulb neurogenesis. Eur. J. Neurosci. 17, 2042–2046.

Brus, M., Meurisse, M., Franceschini, I., Keller, M., Lévy, F., 2010. Evidence for cell proliferation in the sheep brain and its down-regulation by parturition and interactions with the young. Horm. Behav. 58, 737–746.

Callaway, E., 2017. Reprogrammed cells relieve Parkinson's symptoms in trials. Nature.

Carleton, A., Petreanu, L.T., Lansford, R., Alvarez-Buylla, A., Lledo, P.-M., 2003. Becoming a new neuron in the adult olfactory bulb. Nat. Neurosci. 6, 507–518.

Carmichael, S.T., 2005. Rodent models of focal stroke: size, mechanism, and purpose. NeuroRx 2, 396–409.

Chapman, T.R., Barrientos, R.M., Ahrendsen, J.T., Hoover, J.M., Maier, S.F., Patterson, S.L., 2012. Aging and infection reduce expression of specific brain-derived neurotrophic factor mRNAs in hippocampus. Neurobiol. Aging 33, 832.e1–832.e14.

Chen, G., Rajkowska, G., Du, F., Seraji-Bozorgzad, N., Manji, H.K., 2002. Enhancement of hippocampal neurogenesis by lithium. J. Neurochem. 75, 1729–1734.

Chen, Q., Nakajima, A., Choi, S.H., Xiong, X., Sisodia, S.S., Tang, Y.-P., 2008. Adult neurogenesis is functionally associated with AD-like neurodegeneration. Neurobiol. Dis. 29, 316–326.

Chi, L., Gan, L., Luo, C., Lien, L., Liu, R., 2007. Temporal response of neural progenitor cells to disease onset and progression in amyotrophic lateral sclerosis-like transgenic mice. Stem Cells Dev. 16, 579–588.

Chiu, W.-H., Depboylu, C., Hermanns, G., Maurer, L., Windolph, A., Oertel, W.H., Ries, V., Höglinger, G.U., 2015. Long-term treatment with L-DOPA or pramipexole affects adult neurogenesis and corresponding non-motor behavior in a mouse model of Parkinson's disease. Neuropharmacology 95, 367–376.

Cho, K.-O., Lybrand, Z.R., Ito, N., Brulet, R., Tafacory, F., Zhang, L., Good, L., Ure, K., Kernie, S.G., Birnbaum, S.G., et al., 2015. Aberrant hippocampal neurogenesis contributes to epilepsy and associated cognitive decline. Nat. Commun. 6.

Choi, S.H., Veeraraghavalu, K., Lazarov, O., Marler, S., Ransohoff, R.M., Ramirez, J.M., Sisodia, S.S., 2008. Non-cell-autonomous effects of presenilin 1 variants on enrichment-mediated hippocampal progenitor cell proliferation and differentiation. Neuron 59, 568–580.

Chow, V.W., Mattson, M.P., Wong, P.C., Gleichmann, M., 2010. An overview of APP processing enzymes and products. NeuroMolecular Med. 12, 1–12.

Cicchetti, F., Brownell, A.L., Williams, K., Chen, Y.I., Livni, E., Isacson, O., 2002. Neuroinflammation of the nigrostriatal pathway during progressive 6-OHDA dopamine degeneration in rats monitored by immunohistochemistry and PET imaging. Eur. J. Neurosci. 15, 991–998.

Collin, T., Arvidsson, A., Kokaia, Z., Lindvall, O., 2005. Quantitative analysis of the generation of different striatal neuronal subtypes in the adult brain following excitotoxic injury. Exp. Neurol. 195, 71–80.

Cunningham, L.A., Candelario, K., Li, L., 2012. Roles for HIF-1α in neural stem cell function and the regenerative response to stroke. Behav. Brain Res. 227, 410–417.

Curia, G., Longo, D., Biagini, G., Jones, R.S.G., Avoli, M., 2008. The pilocarpine model of temporal lobe epilepsy. J. Neurosci. Methods 172, 143–157.

D'Hooge, R., De Deyn, P.P., 2001. Applications of the Morris water maze in the study of learning and memory. Brain Res. Rev. 36, 60–90.

David, D.J., Samuels, B.A., Rainer, Q., Wang, J.-W., Marsteller, D., Mendez, I., Drew, M., Craig, D.A., Guiard, B.P., Guilloux, J.-P., et al., 2009. Neurogenesis-dependent and -independent effects of fluoxetine in an animal model of anxiety/depression. Neuron 62, 479–493.

De la Rosa-Prieto, C., Saiz-Sanchez, D., Ubeda-Banon, I., Flores-Cuadrado, A., Martinez-Marcos, A., 2016. Neurogenesis, neurodegeneration, interneuron vulnerability, and amyloid-β in the olfactory bulb of APP/PS1 mouse model of Alzheimer's disease. Front. Neurosci. 10.

De Strooper, B., Iwatsubo, T., Wolfe, M.S., 2012. Presenilins and -secretase: structure, function, and role in Alzheimer disease. Cold Spring Harbor Perspect. Med. 2, a006304.

Deleidi, M., Gasser, T., 2013. The role of inflammation in sporadic and familial Parkinson's disease. Cell. Mol. Life Sci. 70, 4259–4273.

Déry, N., Pilgrim, M., Gibala, M., Gillen, J., Wojtowicz, J.M., MacQueen, G., Becker, S., 2013. Adult hippocampal neurogenesis reduces memory interference in humans: opposing effects of aerobic exercise and depression. Front. Neurosci. 7.

Devanand, D.P., Lee, S., Manly, J., Andrews, H., Schupf, N., Masurkar, A., Stern, Y., Mayeux, R., Doty, R.L., 2015. Olfactory identification deficits and increased mortality in the community: olfaction and mortality. Ann. Neurol. 78, 401–411.

Dhanasekaran, D.N., Reddy, E.P., 2008. JNK signaling in apoptosis. Oncogene 27, 6245.

Doeppner, T.R., Hermann, D.M., 2014. Stem cell-based treatments against stroke: observations from human proof-of-concept studies and considerations regarding clinical applicability. Front. Cell. Neurosci. 8.

Doty, R.L., 2012. Olfaction in Parkinson's disease and related disorders. Neurobiol. Dis. 46, 527–552.

Doty, R.L., Shaman, P., Applebaum, S.L., Giberson, R., Siksorski, L., Rosenberg, L., 1984. Smell identification ability: changes with age. Science 226, 1441–1443.

Ducruet, A.F., Zacharia, B.E., Sosunov, S.A., Gigante, P.R., Yeh, M.L., Gorski, J.W., Otten, M.L., Hwang, R.Y., DeRosa, P.A., Hickman, Z.L., et al., 2012. Complement inhibition promotes endogenous neurogenesis and sustained anti-inflammatory neuroprotection following reperfused stroke. PLoS ONE 7, e38664.

Duman, R.S., Malberg, J., Thome, J., 1999. Neural plasticity to stress and antidepressant treatment. Biol. Psychiatry 46, 1181–1191.

Duman, R.S., Li, N., Liu, R.-J., Duric, V., Aghajanian, G., 2012. Signaling pathways underlying the rapid antidepressant actions of ketamine. Neuropharmacology 62, 35–41.

Edwards, L.L., Quigley, E.M., Pfeiffer, R.F., 1992. Gastrointestinal dysfunction in Parkinson's disease: frequency and pathophysiology. Neurology 42, 726–732.

Ekdahl, C.T., Kokaia, Z., Lindvall, O., 2009. Brain inflammation and adult neurogenesis: the dual role of microglia. Neuroscience 158, 1021–1029.

El Khoury, A., Johnson, L., Aberg-Wistedt, A., Stain-Malmgren, R., 2001. Effects of long-term lithium treatment on monoaminergic functions in major depression. Psychiatry Res. 105, 33–44.
Elger, C.E., Helmstaedter, C., Kurthen, M., 2004. Chronic epilepsy and cognition. Lancet Neurol. 3, 663–672.
Encinas, J.M., Sierra, A., 2012. Neural stem cell deforestation as the main force driving the age-related decline in adult hippocampal neurogenesis. Behav. Brain Res. 227, 433–439.
Enwere, E., 2004. Aging results in reduced epidermal growth factor receptor signaling, diminished olfactory neurogenesis, and deficits in fine olfactory discrimination. J. Neurosci. 24, 8354–8365.
Eriksson, P.S., Perfilieva, E., Björk-Eriksson, T., Alborn, A.-M., Nordborg, C., Peterson, D.A., Gage, F.H., 1998. Neurogenesis in the adult human hippocampus. Nat. Med. 4, 1313–1317.
Ernst, A., Alkass, K., Bernard, S., Salehpour, M., Perl, S., Tisdale, J., Possnert, G., Druid, H., Frisén, J., 2014. Neurogenesis in the striatum of the adult human brain. Cell 156, 1072–1083.
Faravelli, I., Riboldi, G., Nizzardo, M., Simone, C., Zanetta, C., Bresolin, N., Comi, G.P., Corti, S., 2014. Stem cell transplantation for amyotrophic lateral sclerosis: therapeutic potential and perspectives on clinical translation. Cell. Mol. Life Sci. 71, 3257–3268.
Fiala, B.A., Joyce, J.N., Greenough, W.T., 1978. Environmental complexity modulates growth of granule cell dendrites in developing but not adult hippocampus of rats. Exp. Neurol. 59 (3), 372–383.
Furuta, M., Bridges, R.S., 2005. Gestation-induced cell proliferation in the rat brain. Brain Res. Dev. Brain Res. 156, 61–66.
Gadadhar, A., Marr, R., Lazarov, O., 2011. Presenilin-1 regulates neural progenitor cell differentiation in the adult brain. J. Neurosci. 31, 2615–2623.
Galán, L., Gómez-Pinedo, U., Guerrero, A., García-Verdugo, J.M., Matías-Guiu, J., 2017. Amyotrophic lateral sclerosis modifies progenitor neural proliferation in adult classic neurogenic brain niches. BMC Neurol. 17.
Galea, L.A., McEwen, B.S., 1999. Sex and seasonal differences in the rate of cell proliferation in the dentate gyrus of adult wild meadow voles. Neuroscience 89, 955–964.
Galve-Roperh, I., Aguado, T., Palazuelos, J., Guzman, M., 2007. The endocannabinoid system and neurogenesis in health and disease. Neuroscientist 13, 109–114.
Gasser, T., Hardy, J., Mizuno, Y., 2011. Milestones in PD genetics. Mov. Disord. 26, 1042–1048.
Ghosal, K., Stathopoulos, A., Pimplikar, S.W., 2010. APP intracellular domain impairs adult neurogenesis in transgenic mice by inducing neuroinflammation. PLoS ONE 5, e11866.
Glasper, E.R., Gould, E., 2013. Sexual experience restores age-related decline in adult neurogenesis and hippocampal function. Hippocampus 23, 303–312.
Goetz, C.G., 2011. The history of Parkinson's disease: early clinical descriptions and neurological therapies. Cold Spring Harbor Perspect. Med. 1, a008862.
Gotts, J.E., Chesselet, M.-F., 2005. Mechanisms of subventricular zone expansion after focal cortical ischemic injury. J. Comp. Neurol. 488, 201–214.
Gould, E., Woolley, C.S., McEwen, B.S., 1990. Short-term glucocorticoid manipulations affect neuronal morphology and survival in the adult dentate gyrus. Neuroscience 37, 367–375.
Gould, E., McEwen, B.S., Tanapat, P., Galea, L.A., Fuchs, E., 1997. Neurogenesis in the dentate gyrus of the adult tree shrew is regulated by psychosocial stress and NMDA receptor activation. J. Neurosci. 17, 2492–2498.
Gould, E., Beylin, A., Tanapat, P., Reeves, A., Shors, T.J., 1999. Learning enhances adult neurogenesis in the hippocampal formation. Nat. Neurosci. 2 (3), 260–265.
Gurney, M.E., Pu, H., Chiu, A.Y., Dal Canto, M.C., Polchow, C.Y., Alexander, D.D., Caliendo, J., Hentati, A., Kwon, Y.W., Deng, H.X., 1994. Motor neuron degeneration in mice that express a human Cu, Zn superoxide dismutase mutation. Science 264, 1772–1775.
Hamilton, L.K., Truong, M.K.V., Bednarczyk, M.R., Aumont, A., Fernandes, K.J.L., 2009. Cellular organization of the central canal ependymal zone, a niche of latent neural stem cells in the adult mammalian spinal cord. Neuroscience 164, 1044–1056.

Haughey, N.J., Nath, A., Chan, S.L., Borchard, A.C., Rao, M.S., Mattson, M.P., 2002a. Disruption of neurogenesis by amyloid β-peptide, and perturbed neural progenitor cell homeostasis, in models of Alzheimer's disease: amyloid-β disrupts neurogenesis. J. Neurochem. 83, 1509–1524.

Haughey, N.J., Liu, D., Nath, A., Borchard, A.C., Mattson, M.P., 2002b. Disruption of neurogenesis in the subventricular zone of adult mice, and in human cortical neuronal precursor cells in culture, by amyloid beta-peptide: implications for the pathogenesis of Alzheimer's disease. NeuroMolecular Med. 1, 125–135.

Hawken, P.A.R., Jorre, T.J.d.S., Rodger, J., Esmaili, T., Blache, D., Martin, G.B., 2009. Rapid induction of cell proliferation in the adult female ungulate brain (Ovis aries) associated with activation of the reproductive axis by exposure to unfamiliar males. Biol. Reprod. 80, 1146–1151.

He, X.J., Nakayama, H., Dong, M., Yamauchi, H., Ueno, M., Uetsuka, K., Doi, K., 2006. Evidence of apoptosis in the subventricular zone and rostral migratory stream in the MPTP mouse model of Parkinson disease. J. Neuropathol. Exp. Neurol. 65, 873–882.

Hoehn, B.D., Palmer, T.D., Steinberg, G.K., 2005. Neurogenesis in rats after focal cerebral ischemia is enhanced by indomethacin. Stroke 36, 2718–2724.

Höglinger, G.U., Rizk, P., Muriel, M.P., Duyckaerts, C., Oertel, W.H., Caille, I., Hirsch, E.C., 2004. Dopamine depletion impairs precursor cell proliferation in Parkinson disease. Nat. Neurosci. 7, 726–735.

Hou, S.-W., Wang, Y.-Q., Xu, M., Shen, D.-H., Wang, J.-J., Huang, F., Yu, Z., Sun, F.-Y., 2008. Functional integration of newly generated neurons into striatum after cerebral ischemia in the adult rat brain. Stroke 39, 2837–2844.

Hsiao, K., Chapman, P., Nilsen, S., Eckman, C., Harigaya, Y., Younkin, S., Yang, F., Cole, G., 1996. Correlative memory deficits, Abeta elevation, and amyloid plaques in transgenic mice. Science 274, 99–102.

Huang, G.-J., Herbert, J., 2005. Serotonin modulates the suppressive effects of corticosterone on proliferating progenitor cells in the dentate gyrus of the hippocampus in the adult rat. Neuropsychopharmacology 30, 231–241.

Huang, Y.-Q., Wu, C., He, X.-F., Wu, D., He, X., Liang, F.-Y., Dai, G.-Y., Pei, Z., Xu, G.-Q., Lan, Y., 2018. Effects of voluntary wheel-running types on hippocampal neurogenesis and spatial cognition in middle-aged mice. Front. Cell. Neurosci. 12.

Iosif, R.E., Ahlenius, H., Ekdahl, C.T., Darsalia, V., Thored, P., Jovinge, S., Kokaia, Z., Lindvall, O., 2008. Suppression of stroke-induced progenitor proliferation in adult subventricular zone by tumor necrosis factor receptor 1. J. Cereb. Blood Flow Metab. 28, 1574–1587.

Jessberger, S., Parent, J.M., 2015. Epilepsy and adult neurogenesis. Cold Spring Harb. Perspect. Biol. a020677.

Jha, S., Rajendran, R., Davda, J., Vaidya, V.A., 2006. Selective serotonin depletion does not regulate hippocampal neurogenesis in the adult rat brain: differential effects of p-chlorophenylalanine and 5,7-dihydroxytryptamine. Brain Res. 1075, 48–59.

Jin, K., Minami, M., Lan, J.Q., Mao, X.O., Batteur, S., Simon, R.P., Greenberg, D.A., 2001. Neurogenesis in dentate subgranular zone and rostral subventricular zone after focal cerebral ischemia in the rat. Proc. Natl. Acad. Sci. U. S. A. 98, 4710–4715.

Jin, K., Peel, A.L., Mao, X.O., Xie, L., Cottrell, B.A., Henshall, D.C., Greenberg, D.A., 2004a. Increased hippocampal neurogenesis in Alzheimer's disease. Proc. Natl. Acad. Sci. 101, 343–347.

Jin, K., Galvan, V., Xie, L., Mao, X.O., Gorostiza, O.F., Bredesen, D.E., Greenberg, D.A., 2004b. Enhanced neurogenesis in Alzheimer's disease transgenic (PDGF-APPSw,Ind) mice. Proc. Natl. Acad. Sci. 101, 13363–13367.

Jin, Y., Raviv, N., Barnett, A., Bambakidis, N.C., Filichia, E., Luo, Y., 2015. The Shh signaling pathway is upregulated in multiple cell types in cortical ischemia and influences the outcome of stroke in an animal model. PLoS ONE 10, e0124657.

Johansson, C.B., Momma, S., Clarke, D.L., Risling, M., Lendahl, U., Frisén, J., 1999. Identification of a neural stem cell in the adult mammalian central nervous system. Cell 96, 25–34.

Kannangara, T.S., Lucero, M.J., Gil-Mohapel, J., Drapala, R.J., Simpson, J.M., Christie, B.R., van Praag, H., 2011. Running reduces stress and enhances cell genesis in aged mice. Neurobiol. Aging 32, 2279–2286.

Karussis, D., Petrou, P., Kassis, I., 2013. Clinical experience with stem cells and other cell therapies in neurological diseases. J. Neurol. Sci. 324, 1–9.

Kaur, A.W., Ackels, T., Kuo, T.-H., Cichy, A., Dey, S., Hays, C., Kateri, M., Logan, D.W., Marton, T.F., Spehr, M., et al., 2014. Murine pheromone proteins constitute a context-dependent combinatorial code governing multiple social behaviors. Cell 157, 676–688.

Kempermann, G., Gast, D., Gage, F.H., 2002. Neuroplasticity in old age: sustained fivefold induction of hippocampal neurogenesis by long-term environmental enrichment. Ann. Neurol. 52, 135–143.

Kessler, R.C., McGonagle, K.A., Zhao, S., Nelson, C.B., Hughes, M., Eshleman, S., Wittchen, H.U., Kendler, K.S., 1994. Lifetime and 12-month prevalence of DSM-III-R psychiatric disorders in the United States. Results from the National Comorbidity Survey. Arch. Gen. Psychiatry 51, 8–19.

Khurana, S., Melacarne, A., Yadak, R., Schouteden, S., Notelaers, T., Pistoni, M., Maes, C., Verfaillie, C.M., 2014. SMAD signaling regulates CXCL12 expression in the bone marrow Niche, affecting homing and mobilization of hematopoietic progenitors: regulation of CXCL12 expression in HSPC Niche. Stem Cells 32, 3012–3022.

Kikuchi, T., Morizane, A., Doi, D., Magotani, H., Onoe, H., Hayashi, T., Mizuma, H., Takara, S., Takahashi, R., Inoue, H., et al., 2017. Human iPS cell-derived dopaminergic neurons function in a primate Parkinson's disease model. Nature 548, 592–596.

Kim, J., Su, S.C., Wang, H., Cheng, A.W., Cassady, J.P., Lodato, M.A., Lengner, C.J., Chung, C.-Y., Dawlaty, M.M., Tsai, L.-H., et al., 2011. Functional integration of dopaminergic neurons directly converted from mouse fibroblasts. Cell Stem Cell 9, 413–419.

Klegeris, A., Giasson, B.I., Zhang, H., Maguire, J., Pelech, S., McGeer, P.L., 2006. Alpha-synuclein and its disease-causing mutants induce ICAM-1 and IL-6 in human astrocytes and astrocytoma cells. FASEB J. 20, 2000–2008.

Klegeris, A., Pelech, S., Giasson, B.I., Maguire, J., Zhang, H., McGeer, E.G., McGeer, P.L., 2008. Alpha-synuclein activates stress signaling protein kinases in THP-1 cells and microglia. Neurobiol. Aging 29, 739–752.

Kobilo, T., Liu, Q.-R., Gandhi, K., Mughal, M., Shaham, Y., van Praag, H., 2011. Running is the neurogenic and neurotrophic stimulus in environmental enrichment. Learn. Mem. 18, 605–609.

Kohl, Z., Winner, B., Ubhi, K., Rockenstein, E., Mante, M., Münch, M., Barlow, C., Carter, T., Masliah, E., Winkler, J., 2012. Fluoxetine rescues impaired hippocampal neurogenesis in a transgenic A53T synuclein mouse model: fluoxetine in A53T synuclein mice. Eur. J. Neurosci. 35, 10–19.

Kordower, J.H., Chu, Y., Hauser, R.A., Freeman, T.B., Olanow, C.W., 2008. Lewy body-like pathology in long-term embryonic nigral transplants in Parkinson's disease. Nat. Med. 14, 504–506.

Kucerova, J., Pistovcakova, J., Vrskova, D., Dusek, L., Sulcova, A., 2012. The effects of methamphetamine self-administration on behavioural sensitization in the olfactory bulbectomy rat model of depression. Int. J. Neuropsychopharmacol. 15, 1503–1511.

Kuhn, H.G., Dickinson-Anson, H., Gage, F.H., 1996. Neurogenesis in the dentate gyrus of the adult rat: age-related decrease of neuronal progenitor proliferation. J. Neurosci. 16, 2027–2033.

Kuhn, H.G., Winkler, J., Kempermann, G., Thal, L.J., Gage, F.H., 1997. Epidermal growth factor and fibroblast growth factor-2 have different effects on neural progenitors in the adult rat brain. J. Neurosci. 17, 5820–5829.

Kwak, K.-A., Lee, S.-P., Yang, J.-Y., Park, Y.-S., 2018. Current perspectives regarding stem cell-based therapy for Alzheimer's disease. Stem Cells Int. 2018, 1–14.

Lee, R.M., Robbins, S.B., 1998. The relationship between social connectedness and anxiety, self-esteem, and social identity. J. Couns. Psychol. 45, 338–345.

Lee, H.-J., Suk, J.-E., Patrick, C., Bae, E.-J., Cho, J.-H., Rho, S., Hwang, D., Masliah, E., Lee, S.-J., 2010. Direct transfer of alpha-synuclein from neuron to astroglia causes inflammatory responses in synucleinopathies. J. Biol. Chem. 285, 9262–9272.

Lee, J.C., Jin, Y., Jin, J., Kang, B.G., Nam, D.-H., Joo, K.M., Cha, C.I., 2011. Functional neural stem cell isolation from brains of adult mutant SOD1 ($SOD1^{G93A}$) transgenic amyotrophic lateral sclerosis (ALS) mice. Neurol. Res. 33, 33–37.

Li, J.-Y., Englund, E., Holton, J.L., Soulet, D., Hagell, P., Lees, A.J., Lashley, T., Quinn, N.P., Rehncrona, S., Björklund, A., et al., 2008. Lewy bodies in grafted neurons in subjects with Parkinson's disease suggest host-to-graft disease propagation. Nat. Med. 14, 501–503.

Li, X., Floriddia, E.M., Toskas, K., Fernandes, K.J.L., Guérout, N., Barnabé-Heider, F., 2016. Regenerative potential of ependymal cells for spinal cord injuries over time. EBioMedicine 13, 55–65.

Lichtenwalner, R.J., Forbes, M.E., Bennett, S.A., Lynch, C.D., Sonntag, W.E., Riddle, D.R., 2001. Intracerebroventricular infusion of insulin-like growth factor-I ameliorates the age-related decline in hippocampal neurogenesis. Neuroscience 107, 603–613.

Lieberwirth, C., Wang, Z., 2012. The social environment and neurogenesis in the adult mammalian brain. Front. Hum. Neurosci. 6.

Lieberwirth, C., Liu, Y., Jia, X., Wang, Z., 2012. Social isolation impairs adult neurogenesis in the limbic system and alters behaviors in female prairie voles. Horm. Behav. 62, 357–366.

Lindvall, O., Kokaia, Z., 2015. Neurogenesis following stroke affecting the adult brain. Cold Spring Harb. Perspect. Biol. 7, a019034.

Liu, Z., Martin, L.J., 2006. The adult neural stem and progenitor cell niche is altered in amyotrophic lateral sclerosis mouse brain. J. Comp. Neurol. 497, 468–488.

Liu, Z., Fan, Y., Won, S.J., Neumann, M., Hu, D., Zhou, L., Weinstein, P.R., Liu, J., 2007. Chronic treatment with minocycline preserves adult new neurons and reduces functional impairment after focal cerebral ischemia. Stroke 38, 146–152.

Liu, F., You, Y., Li, X., Ma, T., Nie, Y., Wei, B., Li, T., Lin, H., Yang, Z., 2009. Brain injury does not alter the intrinsic differentiation potential of adult neuroblasts. J. Neurosci. 29, 5075–5087.

Lu, J., Manaenko, A., Hu, Q., 2017. Targeting adult neurogenesis for poststroke therapy. Stem Cells Int. 2017, 1–10.

Luo, J., Daniels, S.B., Lennington, J.B., Notti, R.Q., Conover, J.C., 2006. The aging neurogenic subventricular zone. Aging Cell 5, 139–152.

Mak, G.K., Enwere, E.K., Gregg, C., Pakarainen, T., Poutanen, M., Huhtaniemi, I., Weiss, S., 2007. Male pheromone-stimulated neurogenesis in the adult female brain: possible role in mating behavior. Nat. Neurosci. 10, 1003–1011.

Malberg, J.E., Eisch, A.J., Nestler, E.J., Duman, R.S., 2000. Chronic antidepressant treatment increases neurogenesis in adult rat hippocampus. J. Neurosci. 20, 9104–9110.

Marlatt, M.W., Lucassen, P.J., van Praag, H., 2010. Comparison of neurogenic effects of fluoxetine, duloxetine and running in mice. Brain Res. 1341, 93–99.

Marlatt, M.W., Potter, M.C., Lucassen, P.J., van Praag, H., 2012. Running throughout middle-age improves memory function, hippocampal neurogenesis, and BDNF levels in female C57BL/6J mice. Dev. Neurobiol. 72, 943–952.

Masliah, E., Rockenstein, E., Veinbergs, I., Mallory, M., Hashimoto, M., Takeda, A., Sagara, Y., Sisk, A., Mucke, L., 2000. Dopaminergic loss and inclusion body formation in alpha-synuclein mice: implications for neurodegenerative disorders. Science 287, 1265–1269.

McEwen, B.S., 1999. Stress and hippocampal plasticity. Annu. Rev. Neurosci. 22, 105–122.

McGeer, P.L., McGeer, E.G., 2008. Glial reactions in Parkinson's disease. Mov. Disord. 23, 474–483.

McGeer, P.L., Itagaki, S., Boyes, B.E., McGeer, E.G., 1988. Reactive microglia are positive for HLA-DR in the substantia nigra of Parkinson's and Alzheimer's disease brains. Neurology 38, 1285–1291.

Meletis, K., Barnabé-Heider, F., Carlén, M., Evergren, E., Tomilin, N., Shupliakov, O., Frisén, J., 2008. Spinal cord injury reveals multilineage differentiation of ependymal cells. PLoS Biol. 6, e182.
Merkle, F.T., Mirzadeh, Z., Alvarez-Buylla, A., 2007. Mosaic organization of neural stem cells in the adult brain. Science 317, 381–384.
Miller, B.R., Hen, R., 2015. The current state of the neurogenic theory of depression and anxiety. Curr. Opin. Neurobiol. 30, 51–58.
Mobley, A.S., Bryant, A.K., Richard, M.B., Brann, J.H., Firestein, S.J., Greer, C.A., 2013. Age-dependent regional changes in the rostral migratory stream. Neurobiol. Aging 34, 1873–1881.
Mohammad, H., Marchisella, F., Ortega-Martinez, S., Hollos, P., Eerola, K., Komulainen, E., Kulesskaya, N., Freemantle, E., Fagerholm, V., Savontous, E., et al., 2018. JNK1 controls adult hippocampal neurogenesis and imposes cell-autonomous control of anxiety behaviour from the neurogenic niche. Mol. Psychiatry 23, 362–374.
Moon, H.Y., Becke, A., Berron, D., Becker, B., Sah, N., Benoni, G., Janke, E., Lubejko, S.T., Greig, N.H., Mattison, J.A., et al., 2016. Running-induced systemic cathepsin B secretion is associated with memory function. Cell Metab. 24, 332–340.
Morris, R.G.M., 1981. Spatial localization does not require the presence of local cues. Learn. Motiv. 12, 239–260.
Murphy, C., Schubert, C.R., Cruickshanks, K.J., Klein, B.E.K., Klein, R., Nondahl, D.M., 2002. Prevalence of olfactory impairment in older adults. JAMA 288, 2307–2312.
Muthukumar, S., Rajesh, D., Saibaba, G., Alagesan, A., Rengarajan, R.L., Archunan, G., 2013. Urinary lipocalin protein in a female rodent with correlation to phases in the estrous cycle: an experimental study accompanied by in silico analysis. PLoS ONE 8, e71357.
Nestler, E.J., Barrot, M., DiLeone, R.J., Eisch, A.J., Gold, S.J., Monteggia, L.M., 2002. Neurobiology of depression. Neuron 34, 13–25.
O'Sullivan, S.S., Johnson, M., Williams, D.R., Revesz, T., Holton, J.L., Lees, A.J., Perry, E.K., 2011. The effect of drug treatment on neurogenesis in Parkinson's disease. Mov. Disord. 26, 45–50.
Ohab, J.J., Fleming, S., Blesch, A., Carmichael, S.T., 2006. A neurovascular Niche for neurogenesis after stroke. J. Neurosci. 26, 13007–13016.
Ohm, T.G., Braak, H., 1987. Olfactory bulb changes in Alzheimer's disease. Acta Neuropathol. 73, 365–369.
Olesen, M.V., Wörtwein, G., Folke, J., Pakkenberg, B., 2017. Electroconvulsive stimulation results in long-term survival of newly generated hippocampal neurons in rats: survival of newly formed neurons in the rat hippocampus. Hippocampus 27, 52–60.
Ormerod, B.K., Lee, T.T.-Y., Galea, L.A.M., 2004. Estradiol enhances neurogenesis in the dentate gyri of adult male meadow voles by increasing the survival of young granule neurons. Neuroscience 128, 645–654.
Parent, J.M., Yu, T.W., Leibowitz, R.T., Geschwind, D.H., Sloviter, R.S., Lowenstein, D.H., 1997. Dentate granule cell neurogenesis is increased by seizures and contributes to aberrant network reorganization in the adult rat hippocampus. J. Neurosci. 17, 3727–3738.
Parent, J.M., Vexler, Z.S., Gong, C., Derugin, N., Ferriero, D.M., 2002. Rat forebrain neurogenesis and striatal neuron replacement after focal stroke. Ann. Neurol. 52, 802–813.
Patrylo, P.R., Tyagi, I., Willingham, A.L., Lee, S., Williamson, A., 2007. Dentate filter function is altered in a proepileptic fashion during aging. Epilepsia 48, 1964–1978.
Petrik, D., Lagace, D.C., Eisch, A.J., 2012. The neurogenesis hypothesis of affective and anxiety disorders: are we mistaking the scaffolding for the building? Neuropharmacology 62, 21–34.
Pinto, J.M., Wroblewski, K.E., Kern, D.W., Schumm, L.P., McClintock, M.K., 2014. Olfactory dysfunction predicts 5-year mortality in older adults. PLoS ONE 9, e107541.
Qiao, L.-Y., Huang, F.-J., Zhao, M., Xie, J.-H., Shi, J., Wang, J., Lin, X.-Z., Zuo, H., Wang, Y.-L., Geng, T.-C., 2014. A two-year follow-up study of cotransplantation with neural stem/progenitor cells and mesenchymal stromal cells in ischemic stroke patients. Cell Transplant. 23, 65–72.

Quarta, E., Bravi, R., Scambi, I., Mariotti, R., Minciacchi, D., 2015. Increased anxiety-like behavior and selective learning impairments are concomitant to loss of hippocampal interneurons in the presymptomatic SOD1(G93A) ALS mouse model: hippocampal changes in SOD1(G93A) mice. J. Comp. Neurol. 523, 1622–1638.
Rinwa, P., Kumar, A., Garg, S., 2013. Suppression of neuroinflammatory and apoptotic signaling cascade by curcumin alone and in combination with piperine in rat model of olfactory bulbectomy induced depression. PLoS ONE 8, e61052.
Roberts, R.C., Force, M., Kung, L., 2002. Dopaminergic synapses in the matrix of the ventrolateral striatum after chronic haloperidol treatment. Synapse 45, 78–85.
Robin, A.M., Zhang, Z.G., Wang, L., Zhang, R.L., Katakowski, M., Zhang, L., Wang, Y., Zhang, C., Chopp, M., 2006. Stromal cell-derived factor 1α mediates neural progenitor cell motility after focal cerebral ischemia. J. Cereb. Blood Flow Metab. 26, 125–134.
Rodríguez, J.J., Jones, V.C., Tabuchi, M., Allan, S.M., Knight, E.M., LaFerla, F.M., Oddo, S., Verkhratsky, A., 2008. Impaired adult neurogenesis in the dentate gyrus of a triple transgenic mouse model of Alzheimer's disease. PLoS ONE 3, e2935.
Rossi, C., Angelucci, A., Costantin, L., Braschi, C., Mazzantini, M., Babbini, F., Fabbri, M.E., Tessarollo, L., Maffei, L., Berardi, N., et al., 2006. Brain-derived neurotrophic factor (BDNF) is required for the enhancement of hippocampal neurogenesis following environmental enrichment. Eur. J. Neurosci. 24, 1850–1856.
Rotunno, M.S., Bosco, D.A., 2013. An emerging role for misfolded wild-type SOD1 in sporadic ALS pathogenesis. Front. Cell. Neurosci. 7.
Sah, N., Peterson, B.D., Lubejko, S.T., Vivar, C., van Praag, H., 2017. Running reorganizes the circuitry of one-week-old adult-born hippocampal neurons. Sci. Rep. 7.
Saiz-Sanchez, D., De La Rosa-Prieto, C., Ubeda-Bañon, I., Martinez-Marcos, A., 2013. Interneurons and beta-amyloid in the olfactory bulb, anterior olfactory nucleus and olfactory tubercle in APPxPS1 transgenic mice model of Alzheimer's disease: interneurons and Aβ in APPx PS1 mouse. Anat. Rec. 296, 1413–1423.
Santarelli, L., 2003. Requirement of hippocampal neurogenesis for the behavioral effects of antidepressants. Science 301, 805–809.
Sapolsky, n., 1996. Stress, glucocorticoids, and damage to the nervous system: the current state of confusion. Stress 1, 1–19.
Sheline, Y.I., Wang, P.W., Gado, M.H., Csernansky, J.G., Vannier, M.W., 1996. Hippocampal atrophy in recurrent major depression. Proc. Natl. Acad. Sci. U. S. A. 93, 3908–3913.
Shingo, T., 2003. Pregnancy-stimulated neurogenesis in the adult female forebrain mediated by prolactin. Science 299, 117–120.
Shruster, A., Ben-Zur, T., Melamed, E., Offen, D., 2012. Wnt signaling enhances neurogenesis and improves neurological function after focal ischemic injury. PLoS ONE 7, e40843.
Slotkin, T.A., Seidler, F.J., 2006. Cholinergic receptor subtypes in the olfactory bulbectomy model of depression. Brain Res. Bull. 68, 341–345.
Smith, M.T., Pencea, V., Wang, Z., Luskin, M.B., Insel, T.R., 2001. Increased number of BrdU-labeled neurons in the rostral migratory stream of the estrous prairie vole. Horm. Behav. 39, 11–21.
Snyder, J.S., Soumier, A., Brewer, M., Pickel, J., Cameron, H.A., 2011. Adult hippocampal neurogenesis buffers stress responses and depressive behaviour. Nature 476, 458–461.
Sohrabi, H.R., Bates, K.A., Weinborn, M.G., Johnston, A.N.B., Bahramian, A., Taddei, K., Laws, S.M., Rodrigues, M., Morici, M., Howard, M., et al., 2012. Olfactory discrimination predicts cognitive decline among community-dwelling older adults. Transl. Psychiatry 2, e118.
Song, C., Leonard, B.E., 2005. The olfactory bulbectomised rat as a model of depression. Neurosci. Biobehav. Rev. 29, 627–647.
Sorrells, S.F., Paredes, M.F., Cebrian-Silla, A., Sandoval, K., Qi, D., Kelley, K.W., James, D., Mayer, S., Chang, J., Auguste, K.I., et al., 2018. Human hippocampal neurogenesis drops sharply in children to undetectable levels in adults. Nature 555, 377–381.
Spritzer, M.D., Ibler, E., Inglis, W., Curtis, M.G., 2011. Testosterone and social isolation influence adult neurogenesis in the dentate gyrus of male rats. Neuroscience 195, 180–190.

Stevens, J.C., Cain, W.S., Demarque, A., Ruthruff, A.M., 1991. On the discrimination of missing ingredients: aging and salt flavor. Appetite 16, 129–140.

Stroth, S., Hille, K., Spitzer, M., Reinhardt, R., 2009. Aerobic endurance exercise benefits memory and affect in young adults. Neuropsychol. Rehabil. 19, 223–243.

Sulzer, D., 2007. Multiple hit hypotheses for dopamine neuron loss in Parkinson's disease. Trends Neurosci. 30, 244–250.

Suzuki, Y., Yamamoto, S., Umegaki, H., Onishi, J., Mogi, N., Fujishiro, H., Iguchi, A., 2004. Smell identification test as an indicator for cognitive impairment in Alzheimer's disease. Int. J. Geriatr. Psychiatry 19, 727–733.

Talmi, M., Carlier, E., Soumireu-Mourat, B., 1993. Similar effects of aging and corticosterone treatment on mouse hippocampal function. Neurobiol. Aging 14, 239–244.

Tanapat, P., Hastings, N.B., Reeves, A.J., Gould, E., 1999. Estrogen stimulates a transient increase in the number of new neurons in the dentate gyrus of the adult female rat. J. Neurosci. 19, 5792–5801.

Tazaki, A., Tanaka, E.M., Fei, J.-F., 2017. Salamander spinal cord regeneration: the ultimate positive control in vertebrate spinal cord regeneration. Dev. Biol. 432, 63–71.

Teicher, M.H., Stewart, W.B., Kauer, J.S., Shepherd, G.M., 1980. Suckling pheromone stimulation of a modified glomerular region in the developing rat olfactory bulb revealed by the 2-deoxyglucose method. Brain Res. 194, 530–535.

Teng, Y.D., Benn, S.C., Kalkanis, S.N., Shefner, J.M., Onario, R.C., Cheng, B., Lachyankar, M.B., Marconi, M., Li, J., Yu, D., et al., 2012. Multimodal actions of neural stem cells in a mouse model of ALS: a meta-analysis. Sci. Transl. Med. 4, 165ra164.

Teramoto, T., Qiu, J., Plumier, J.-C., Moskowitz, M.A., 2003. EGF amplifies the replacement of parvalbumin-expressing striatal interneurons after ischemia. J. Clin. Invest. 111, 1125–1132.

Thibault, O., Hadley, R., Landfield, P.W., 2001. Elevated postsynaptic $[Ca^{2+}]_i$ and L-type calcium channel activity in aged hippocampal neurons: relationship to impaired synaptic plasticity. J. Neurosci. 21, 9744–9756.

Ticozzi, N., Tiloca, C., Calini, D., Gagliardi, S., Altieri, A., Colombrita, C., Cereda, C., Ratti, A., Pezzoli, G., Borroni, B., et al., 2014. C9orf72 repeat expansions are restricted to the ALS-FTD spectrum. Neurobiol. Aging 35, 936.e13–936.e17.

Tonchev, A.B., Yamashima, T., 2006. Differential neurogenic potential of progenitor cells in dentate gyrus and CA1 sector of the postischemic adult monkey hippocampus. Exp. Neurol. 198, 101–113.

van den Berge, S.A., van Strien, M.E., Korecka, J.A., Dijkstra, A.A., Sluijs, J.A., Kooijman, L., Eggers, R., De Filippis, L., Vescovi, A.L., Verhaagen, J., et al., 2011. The proliferative capacity of the subventricular zone is maintained in the parkinsonian brain. Brain 134, 3249–3263.

van Praag, H., Christie, B.R., Sejnowski, T.J., Gage, F.H., 1999. Running enhances neurogenesis, learning, and long-term potentiation in mice. Proc. Natl. Acad. Sci. U. S. A. 96, 13427–13431.

Velayudhan, L., 2015. Smell identification function and Alzheimer's disease: a selective review. Curr. Opin. Psychiatry 28, 173–179.

Verret, L., Jankowsky, J.L., Xu, G.M., Borchelt, D.R., Rampon, C., 2007. Alzheimer's-type amyloidosis in transgenic mice impairs survival of newborn neurons derived from adult hippocampal neurogenesis. J. Neurosci. 27, 6771–6780.

Villeda, S.A., Luo, J., Mosher, K.I., Zou, B., Britschgi, M., Bieri, G., Stan, T.M., Fainberg, N., Ding, Z., Eggel, A., et al., 2011. The ageing systemic milieu negatively regulates neurogenesis and cognitive function. Nature 477, 90–94.

Vitalis, T., Ansorge, M.S., Dayer, A.G., 2013. Serotonin homeostasis and serotonin receptors as actors of cortical construction: special attention to the 5-HT3A and 5-HT6 receptor subtypes. Front. Cell. Neurosci. 7.

Vivar, C., Peterson, B.D., van Praag, H., 2016. Running rewires the neuronal network of adult-born dentate granule cells. NeuroImage 131, 29–41.

Voltarelli, J.C., Couri, C.E.B., Stracieri, A.B.P.L., Oliveira, M.C., Moraes, D.A., Pieroni, F., Coutinho, M., Malmegrim, K.C.R., Foss-Freitas, M.C., Simões, B.P., et al., 2007. Autologous nonmyeloablative hematopoietic stem cell transplantation in newly diagnosed type 1 diabetes mellitus. JAMA 297, 1568.

Walsh, R.N., Budtz-Olsen, O.E., Penny, J.E., Cummins, R.A., 1969. The effects of environmental complexity on the histology of the rat hippocampus. J. Comp. Neurol. 137 (3), 361–366.

Wang, C., Liu, F., Liu, Y.-Y., Zhao, C.-H., You, Y., Wang, L., Zhang, J., Wei, B., Ma, T., Zhang, Q., et al., 2011. Identification and characterization of neuroblasts in the subventricular zone and rostral migratory stream of the adult human brain. Cell Res. 21, 1534–1550.

Wang, S., Okun, M.S., Suslov, O., Zheng, T., McFarland, N.R., Vedam-Mai, V., Foote, K.D., Roper, S.N., Yachnis, A.T., Siebzehnrubl, F.A., et al., 2012a. Neurogenic potential of progenitor cells isolated from postmortem human Parkinsonian brains. Brain Res. 1464, 61–72.

Wang, Z., Andrade, N., Torp, M., Wattananit, S., Arvidsson, A., Kokaia, Z., Jørgensen, J.R., Lindvall, O., 2012b. Meteorin is a chemokinetic factor in neuroblast migration and promotes stroke-induced striatal neurogenesis. J. Cereb. Blood Flow Metab. 32, 387–398.

Wang, Y., Ji, X., Leak, R.K., Chen, F., Cao, G., 2017. Stem cell therapies in age-related neurodegenerative diseases and stroke. Ageing Res. Rev. 34, 39–50.

Wilson, R.S., Arnold, S.E., Schneider, J.A., Tang, Y., Bennett, D.A., 2007. The relationship between cerebral Alzheimer's disease pathology and odour identification in old age. J. Neurol. Neurosurg. Psychiatry 78, 30–35.

Winchester, C.L., Ohzeki, H., Vouyiouklis, D.A., Thompson, R., Penninger, J.M., Yamagami, K., Norrie, J.D., Hunter, R., Pratt, J.A., Morris, B.J., 2012. Converging evidence that sequence variations in the novel candidate gene MAP2K7 (MKK7) are functionally associated with schizophrenia. Hum. Mol. Genet. 21, 4910–4921.

Winner, B., Geyer, M., Couillard-Despres, S., Aigner, R., Bogdahn, U., Aigner, L., Kuhn, G., Winkler, J., 2006. Striatal deafferentation increases dopaminergic neurogenesis in the adult olfactory bulb. Exp. Neurol. 197, 113–121.

Winner, B., Kohl, Z., Gage, F.H., 2011. Neurodegenerative disease and adult neurogenesis: neurodegenerative disease and adult neurogenesis. Eur. J. Neurosci. 33, 1139–1151.

Woolley, C.S., Gould, E., McEwen, B.S., 1990. Exposure to excess glucocorticoids alters dendritic morphology of adult hippocampal pyramidal neurons. Brain Res. 531, 225–231.

Worlitzer, M.M., Bunk, E.C., Hemmer, K., Schwamborn, J.C., 2012. Anti-inflammatory treatment induced regenerative oligodendrogenesis in parkinsonian mice. Stem Cell Res Ther 3, 33.

Yan, W., Wilson, C.C., Haring, J.H., 1997. 5-HT1a receptors mediate the neurotrophic effect of serotonin on developing dentate granule cells. Brain Res. Dev. Brain Res. 98, 185–190.

Yang, C.-P., Gilley, J.A., Zhang, G., Kernie, S.G., 2011. ApoE is required for maintenance of the dentate gyrus neural progenitor pool. Development 138, 4351–4362.

Zhang, R., Zhang, Z., Zhang, C., Zhang, L., Robin, A., Wang, Y., Lu, M., Chopp, M., 2004. Stroke transiently increases subventricular zone cell division from asymmetric to symmetric and increases neuronal differentiation in the adult rat. J. Neurosci. 24, 5810–5815.

Zhang, J., Walker, W.B., Wang, G., 2015. Pheromone reception in moths: from molecules to behaviors. Prog. Mol. Biol. Transl. Sci. 130, 109–128.

Zhang, C.-L., Huang, T., Wu, B.-L., He, W.-X., Liu, D., 2017. Stem cells in cancer therapy: opportunities and challenges. Oncotarget 8.

Glossary

Achaete-scute homolog 1 (Ascl1) A proneural gene that controls the specification of neural progenitors and later steps of differentiation and neurite outgrowth.

Action potential A pulse of increased voltage that travels rapidly through a neuron.

α-Amino-3-hydroxy-5-methyl-4-isoxazolepropionic acid (AMPA) An amino acid whose structure is similar to glutamate. AMPA binds to a receptor that activates glutamate-gated ion channels, referred to as AMPA receptors.

Amyloidogenic Producing a starch-like protein that can form aggregates or deposits.

Angiogenesis The production of new blood vessels.

Asymmetric cell division A stem cell divides to produce one identical copy of itself (self-renewal) and one cell that is different from the mother cell, typically more committed to becoming a specific cell type than the original.

β-Integrin A member of the integrin family, a large group of transmembrane receptors that adhere a cell's cytoskeleton to the extracellular matrix or other cells.

Brain lipid binding protein (BLBP) Contributes to fatty acid uptake, transport and metabolism.

Calbindin An intracellular protein and member of the EF-hand proteins (transcription factor category) with high affinity for calcium. Contributes to calcium buffering and synaptic plasticity.

Calretinin A calcium binding protein and member of the EF-hand proteins. Contributes to calcium buffering and modulates intrinsic neuronal excitability.

CD markers Clusters of differentiation markers. A defined set of antibodies known to label a specific antigen.

Cellular reprogramming The ability to alter the fate of a cell by changing the genes it expresses.

Centrifugal Axon projections from higher, more central processing areas of the brain.

Chemoattraction Movement of a cell towards a detected substance or molecule.

Chemokines Proteins arranged in a gradient on nonreceptor sites that attracts cells expressing receptors for the chemokine that induces the cells to mobilize or migrate.

Chemorepulsion Movement of a cell away from a detected substance or molecule.

Chemosensory Relating to the chemosensory systems, taste and smell.

Chimeric An organism whose cells have at least two distinct genotypes. One way of producing chimeric mice is to add iPSCs to a blastocyst.

Chromatin A tightly condensed complex of DNA and histone proteins.

Conditioned media Media taken from a dish with a specific cell type growing in it. The cells release undefined biologically active factors.

Cytokines Hormone-like compounds secreted by a broad range of cells that have an effect on the cell of origin (autocrine), surrounding cells (paracrine) or a biological system (endocrine). A given cytokine can be produced by more than one cell type. For example, interleukin 6 (IL-6)-like cytokine was originally shown to play a role in B cell maturation in the immune system; however, it is also involved in neurogenesis and mature neuron and glial responses (Erta et al., 2012; Roldán and Brieva, 1991).

Dedifferentiate To reverse the state of a cell from a determined developmental fate with specialized characteristics to an undifferentiated multipotent or pluripotent state. The loss of specialized cellular characteristics; to revert back to an earlier stage of development.

Dendritic spines A small protrusion from a dendrite that contains all the elements of a postsynaptic terminal. Each spine typically forms a synapse with a single axon.

Dendrodentritic Synapses that form between neuronal dendrites.

Depolarization A more positive change in potential between the inside and outside of the plasma membrane of a nerve cell due to migration of sodium ions to the intracellular side.

Differentiate A stem or progenitor cell divides and one or both daughter cells become a more specialized cell type.

Diffusion The movement of molecules from a region of high concentration to regions of low concentration.

Doublecortin (DCX) A microtubule-associated protein that stabilizes and causes bundling of microtubules. A common marker of migratory and immature neurons.

Electrostatic pressure The attractive force between charged particles with opposite signs or the repulsive force between two charged particles with the same sign.

Embryonic stem cell Pluripotent stem cells present during the blastocyst stage of embryo development. They are self-renewing cells and have the potential to become any cell in the body.

Endonuclease An enzyme that cleaves the phosphodiester bond between two DNA nucleotides typically at a specific sequence unless it is a nonspecific endonuclease.

Endosteum A type of connective tissue made up of osteoblasts, interspersed with osteoclasts that line the inner surface of the bony tissue surrounding the medullary cavity.

Engraft The process in which HSCs are recruited to the vascular HSC niche where they begin to proliferate.

Engram Persistent changes in synaptic strength via LTP or LTD of preexisting connections as a mechanism for memory formation.

Epigenetics The regulation of gene transcription by elements around the DNA but not of the DNA. Elements that produce "open" chromatin, histone modifications and the interaction of transcription factors with enhancers and promoters are epigenetic in nature.

Extrachromosomal DNA that is located outside of the chromosomes. In the case of viral DNA, it remains in the cytoplasm where it can replicate independently.

Fibroblasts The most common connective tissue cell in animals that produces extracellular matrix and collagen.

Gamma-aminobutyric acid (GABA) Neurons that express the neurotransmitter GABA are called GABAergic. There are two types of GABA receptors, $GABA_A$ and $GABA_B$. $GABA_A$ receptors gate ionotropic Cl– channels and are sensitive to bicuculline antagonism. $GABA_B$ receptors are G protein-coupled receptors that conduct an inward K+ current and can be blocked by baclofen.

Generative asymmetric cell division A stem cell produces one identical copy of itself (self-renewal) and one cell that is specialized. In the brain this occurs over a long period to generate neural cells (neurons and glia).

Generative symmetric cell division At the end of development stem cells may switch from asymmetric to symmetric cell division and generate two specialized cells, thus reducing the stem cell population.

Germ cell Sperm or ovum containing one set of chromosomes after meiosis that can combine with a germ cell from the opposite sex resulting in an embryo.

Germ layers Early cell specifications of the embryo that give rise to explicit tissue types in the body.

Glial fibrillary acidic protein (GFAP) A member of the intermediate filament family expressed in astroglial cells. Most likely involved in controlling the shape, movement and function of astrocytes.

Glomerulus A cluster of axonal synaptic terminals and dendritic endings in the OB.

Glycoproteins A class of proteins that has carbohydrate groups, called oligosaccharide chains, attached to amino acid side chains.

Haploid A cell that has half the diploid number of chromosomes.

Homing When HSCs are induced by the chemokine CXCL12 to return from circulating in the peripheral blood to a hematopoietic niche through endothelial sinusoids.

Hyperpolarization A more negative change in potential between the inside and outside of the plasma membrane of a nerve cell due to migration of potassium ions to the extracellular side.

Hypokinesia Partial or complete loss of muscle movement.

Induced Pluripotent Stem Cells (iPSCs) A pluripotent stem cell artificially derived from somatic cells by inducing the expression of specific genes.

Integral membrane proteins A protein that is structurally or functionally a permanent part of the membrane, or firmly anchored there.

Ischemia A restriction in blood supply resulting in a loss of oxygen, and death to the cells.

Keratinocytes Cells of the epidermis that produce keratin.

Kinase An enzyme responsible for phosphorylation of a substrate through the transfer of a phosphate group from ATP.

Lamellipodia Latin for "thin sheet" and "foot." Webbing-like sheets of actin that extend forward or retract like a wave; a morphological property of migrating cells.

Lateral inhibition The phenomenon that occurs when a neuron undergoes an action potential and releases inhibitory neurotransmitter onto neighboring cells.

Locus A region of DNA consisting of several millions of base pairs.

Membrane potential The voltage measured inside a cell relative to the voltage just outside it.

Mesenchymal cells Arising from the germ layer called the mesenchyme.

MicroRNA A small noncoding RNA fragment of 19–22 nucleotides that prevents expression of a protein by binding to the messenger RNA that would have produced it, resulting in the destruction of the double-strand of RNA.

Mobilization A process that induces HSCs to enter the blood stream from the niche. Granulocyte-colony stimulating factor normally induces mobilization.

Motor protein A class of proteins that move along the cytoskeleton by using ATP energy that causes a conformational change that translates into movement.

Multipotent The potential to become any cell type *within* a germ layer, organ or tissue.

***N*-Methyl-D-aspartate (NMDA)** An amino acid whose structure is similar to glutamate. NMDA binds to a receptor that activates glutamate-gated ion channels, referred to as NMDA receptors.

Nestin An intermediate filament protein expressed in dividing cells; a marker of self-renewal. Expressed in the majority of mitotically active CNS progenitors that give rise to neurons and glia.

NeuN A neuronal nuclear antigen commonly used as a biomarker for neurons. A homologue to sex-determining genes in *Caenorhabditis elegans*.

Neural cells Encompasses neuronal and glial cells.

Neural progenitor cell (NPC) Similar to transit-amplifying cells ***(Chapter 2). These are cells that proliferate for a finite period of time, giving rise to and then differentiating into neuronal or glial cell types.

NeuroD A proneural basic helix-loop-helix (bHLH) transcription factor that induces neuronal differentiation.

Neurotransmitter A chemical released from the synaptic terminal that diffuses across the synaptic cleft that confers an excitatory or inhibitory potential to the postsynaptic neuron.

Niche A specialized microenvironment in which stem cells reside. The niche provides a location for stem cells and/or factors that influence the fate of a dividing stem cell. For example, neighboring cells may secrete factors that increase or decrease stem cell proliferation.

Nucleosome The smallest unit of chromatin consisting of DNA wrapped two times (~147 base pairs) around a core of eight proteins called histones.

Open-label A clinical trial in which both the physicians and the patients know if they are in the treatment or control group.

Oxidative stress An imbalance between the production of reactive oxygen species (free radicals) and antioxidant defenses. Inflammation can induce oxidative stress.

Parabiosis "Living beside." The anatomical union of two organisms.

Parvalbumin A calcium binding protein structurally related to calmodulin, with three EF-hand motifs. It is found in fast-firing nerve and muscle cells.

Penumbra The area surrounding an ischemic event.

Pluripotent The potential to become any cell type in the body.

Polysialic neural cell adhesion molecule (PSA-NCAM) The polysialylated form of the cell surface glycoprotein, neural cell adhesion molecule. A marker of developing and migratory neurons.

Primary neurulation A process in vertebrate embryos in which a segment of the developing ectoderm transforms into neuroectoderm or neural plate, followed by a series of proliferation and folding that results in the neural tube.

Primordial germ cells The cells of the embryo that develop into stem cells that produce germ cells, sperm or ovum.

Proliferative symmetric cell division A stem cell divides to produce two identical copies of itself.

Prospero-related homeobox 1 (Prox-1) Promotes self-renewal of embryonic NSCs, neurogenesis and inhibits astrogliogenesis. Suppresses cell cycle progression via a direct action in basic components of the cell cycle machinery.

Quiescence A period in which stem cells do not divide, a state that sustains the integrity of stem cells.

Radial glial cell (RGC) Radially oriented, bipolar cells in the developing vertebrate cortex that have a basal process that persists during cell division to support the migration of new born neurons. RGCs express astrocytic markers but retain expression of the neuroepithelial marker, intermediate filament Nestin. RGCs give rise to neurons, oligodendrocytes and astrocytes, and are the precursor of Type B astrocytes in the SVZ.

Remodel To cause a cell to undergo morphological change.

Reticular cells A type of fibroblast with long processes that extend to other cell types or other reticular cells creating structural support within a tissue.

Self-renewal When stem cells divide to produce one or two new copies of the original.

Solenoid A unit of chromatin consisting of six nucleosomes that form a hexagon-like structure that contributes to the compact nature of DNA within the cell's nucleus.

Somatic cell Adult or terminally differentiated cells of the body containing a full set of chromosomes. Not a reproductive cell.

Somatic stem cells A multi or unipotent stem cell found in adult tissues or organs throughout the body that provides lifelong replacement, replenishment or repair of specific cells.

Sox2 A transcription factor that maintains pluripotency and inhibits neuronal differentiation.

Symmetric cell division (self-renewal) When a stem cell divides to produce two new copies of itself. This is a defining feature of stem cells.

T-box brain protein 2 (Tbr2) A transcription factor also known as Eomesodermin. Tbr2 is required for establishing neuronal progenitors in the adult SGZ.

Teratocarcinoma A combination of two terms, teratoma and embryonic carcinoma. It is a malignant germ cell growth in the testes in which differentiated cells from all three germ layers can be found. In vivo it has the potential to metastasize and invade local tissues. However, ESC-derived masses are typically benign (Lensch and Ince, 2007).

TGFβ receptors Transforming growth factor beta is a ligand that binds to any of four different splice versions of a transmembrane serine/threonine kinase (Trk) receptor.

Threshold of excitation The potential at which a depolarization must reach for an action potential to occur.

Totipotent A cell with the ability to produce every cell in the body *and* the placenta. Totipotency is the initial state of a fertilized embryo.

Transdifferentiation The regulation of transcription factors to reprogram a cell from one fate to another, specifically involving a lineage conversion. For example, the conversion of a mesodermal cell to an ectodermal cell.

Transduce The transfer of genetic material from one organism to another.

Transit-amplifying cells Cells that arise directly from stem cells that proliferate for a finite period of time and then differentiate.

Transcription factors Proteins that bind to or interact with DNA to produce RNA. The part of the transcription factor that interacts directly with the DNA is called the homeodomain, typically about 60 amino acids long. Transcription factors have several conserved structural motifs that facilitate interaction with DNA such as the basic helix-loop-helix motif.

Tuj1 A neuron-specific Class III β-tubulin that is present in intermediate neural progenitors and immature granule cells.

Tyrosine hydroxylase (TH) The molecule that is the precursor to dopamine. Often used as a marker of dopaminergic neurons in histology studies.

Unipotent The potential to produce a single cell type in addition to its own self-renewal.

Vector A vehicle, such as a virus, used to transfer genetic material to a target cell.

Index

Note: Page numbers followed by *f* indicate figures, *t* indicate tables, and *b* indicate boxes.

T

U

Kolibri 360 GmbH
Ferdinand-Jühlke-Straße 7
99095 Erfurt, Deutschland
produktsicherheit@kolibri360.de